includes a NEW chapter on

EARTH SYSTEMS & CYCLES

journey to the center of understanding

Understanding Earth, 2/e Overview

Introduction	3
New Content	4
New Instructional Tools	6
Biography and Interview with Author Frank Press	10
Biography and Interview with Author Raymond Siever	11
The Supplements for *Understanding Earth*, 2/e	
CD-ROM	12
Press Releases	14
Understanding Earth Web Site	14
Overhead Transparency Set	15
Slide Set with Lecture Notes	15
Test Bank	16
Study Guide	16
Instructor's Resource Manual	16

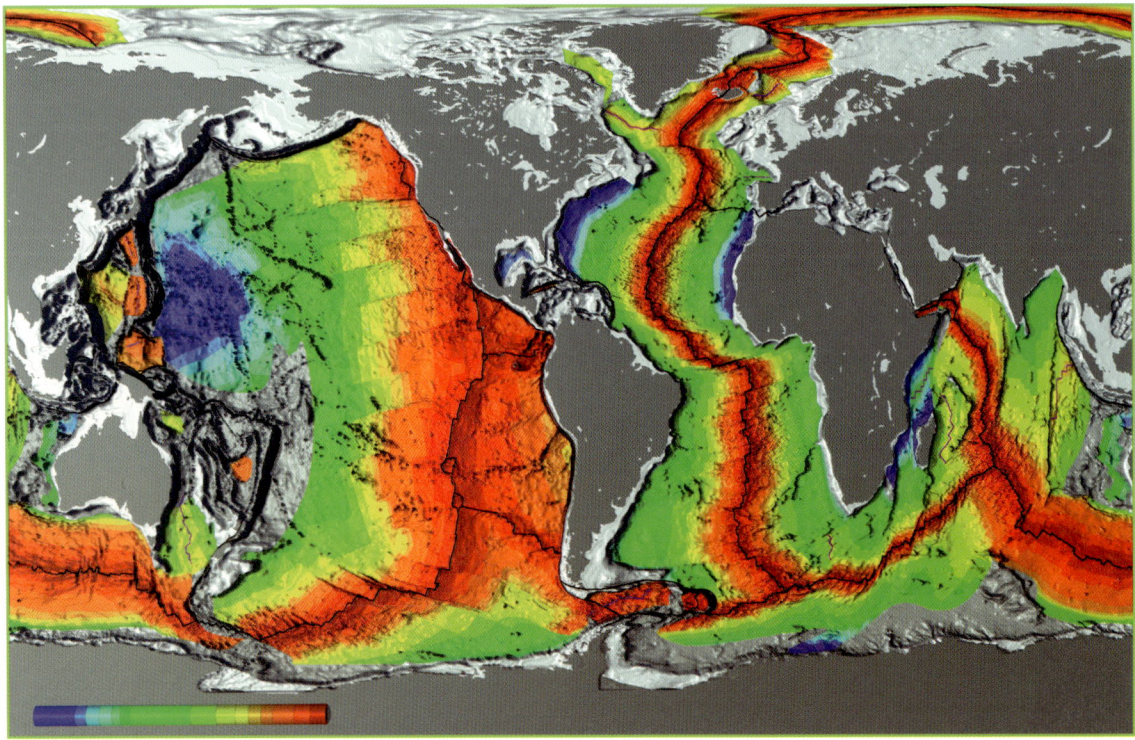

Figure 20.11 This digital image reflects data collected all over the world to interpret the age of the ocean floor. Age color overlay image is by R. Dietmar Meuller, University of Sydney, Australia. Combined age relief images by Peter W. Sloss. **NOAA/NESDIS/NGDC**

THE PLATES OF EARTH'S LITHOSPHERE

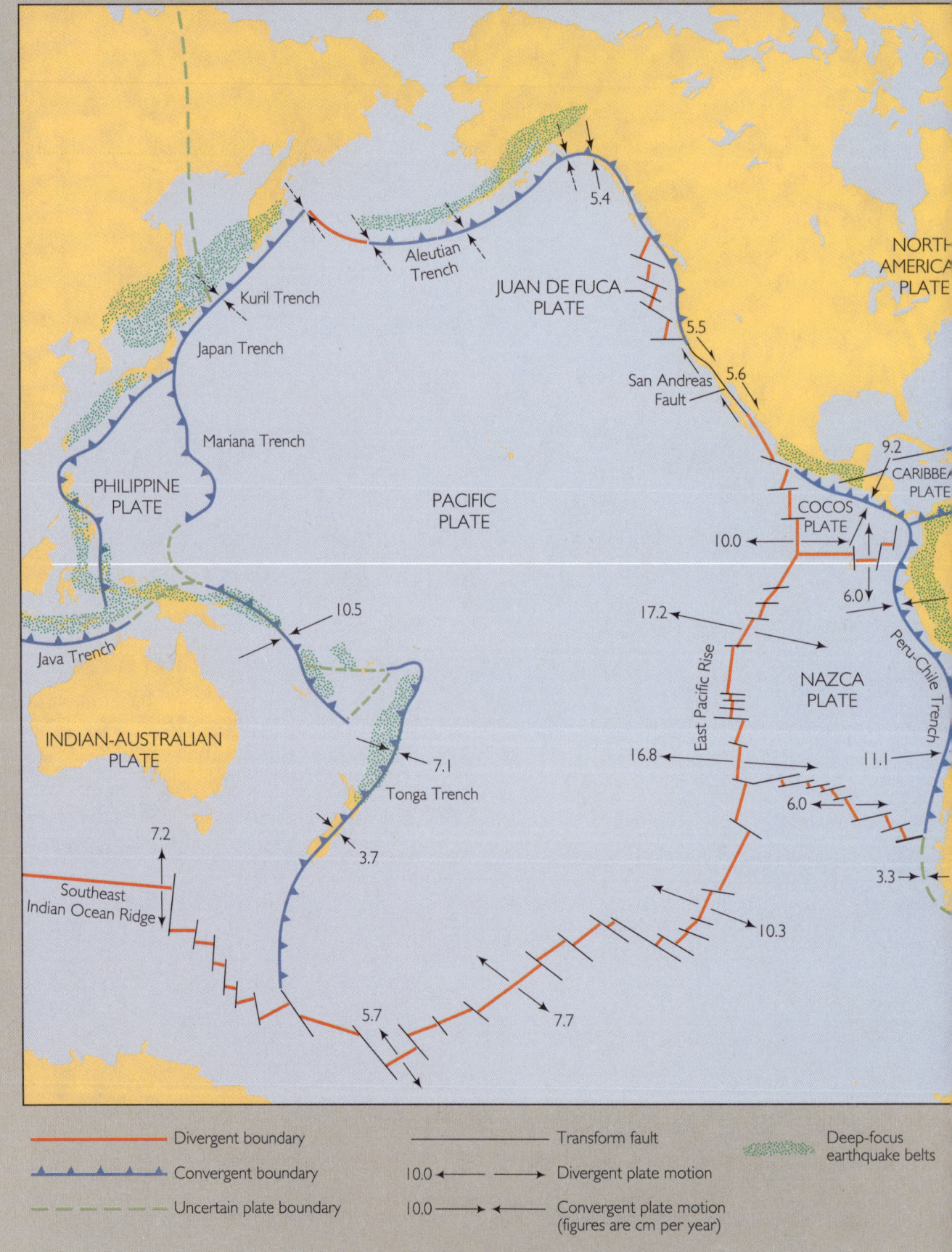

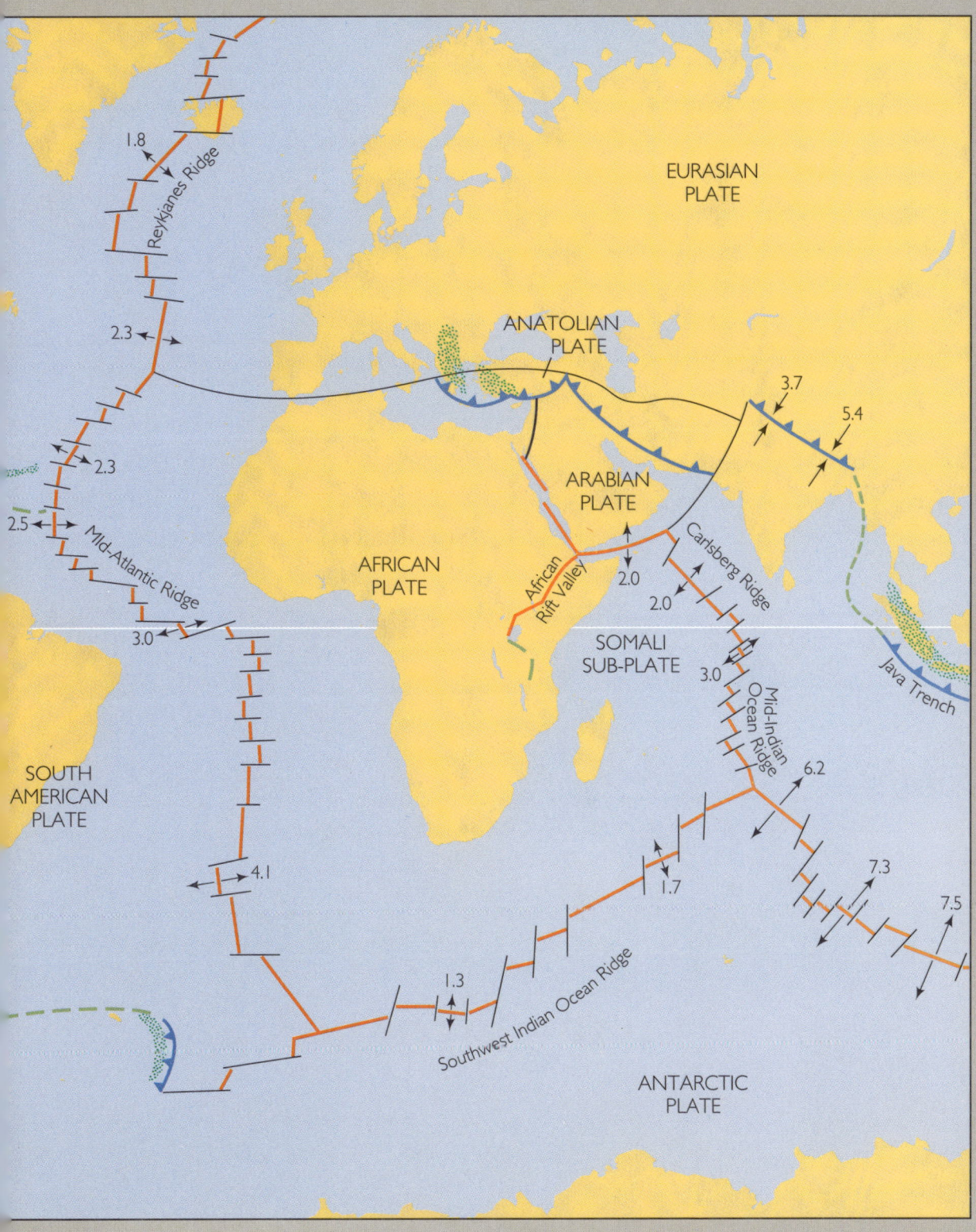

After J. Francheteau, "The Oceanic Crust." Copyright ©1983 by Scientific American, Inc. All rights reserved.
Plate motions from work of C. DeMets, R. G. Gordon, D. F. Argus, S. Stein, Model NUVEL-1, 1990.

Introduction

This preview booklet will be your guide to the second edition of **Understanding Earth**, the Number #1 physical geology textbook in the country. Here you will find the Preface, Chapters 1 through 5, and a Table of Contents for the entire book. Preceding these is a 16-page overview, which presents the distinctive features of this popular textbook and highlights what's new to this edition. Each chapter reflects the newest research findings and incorporates recent geologically significant events. The new Chapter 24, "Earth Systems and Cycles," weaves together the lessons of the preceding twenty-three, reinforcing the systems approach that informs the book from the beginning. More than half the photographs are new in this edition.

What hasn't changed is that the renowned and respected author team—Frank Press of the Carnegie Institution in Washington, D.C.'s Department of Terrestrial Magnetism, and Raymond Siever of Harvard University's Department of Earth and Planetary Sciences—is still vitally interested in engaging the intellect and imagination of today's students. Guided by the authoritative voices of Press and Siever, your students will experience geology—in all its plate-sliding, volcano-erupting, mountain-forming grandeur—as the dynamic science you know it to be. In preparing this text for introductory students, Press and Siever are not only enriching students' lives by introducing them to this exciting field of study, they are preparing tomorrow's decision makers to confront such vital concerns as resource development, waste disposal, environmental protection, and land use. After all, stewardship of humanity's most precious resource—Earth—will soon be in their hands.

*If you would like a free copy of the new edition of **Understanding Earth**, please fill out the order card on page 15 of this overview.*

Figure 5.26 A sulfur-encrusted fumarole on Sierra Negra volcano in the Galápagos Islands (© Christian Grzimek/Photo Researchers OKAPIA)

New Content

Figure 22.14 Solar panels and a villager in Nepal (© Ned Gillette/Stock Market)

NEW CHAPTER

Threaded through the entire book, the theme of Earth systems and cycles culminates in a new final chapter, Chapter 24. Here, the authors address the interdependence of the Earth's interior, crust, atmosphere, hydrosphere, and biosphere. In describing geochemical cycles, they discuss reservoirs, fluxes, and residence time using as examples the calcium and carbon cycles. They relate the mass extinctions of the Permian-Triassic and the Cretaceous-Tertiary boundaries to possible geological causes of climate change. They also look at the implications of human activity for global change and what that will mean for the survival of species, including our own.

A SAMPLING OF NEW MATERIAL

✔ **Chapter 1** includes an expanded treatment of the scientific method, including a new boxed feature. (See pages 4, 22, and 23 in this preview book.)

✔ **Chapter 4** includes underplating at subduction zones (see page 86) and a new diagram of magmas and hot rock at mid-ocean ridges (see page 99).

✔ **Chapter 5** contains new material on edifice collapse—a newly recognized and extremely catastrophic feature of volcanism (see page 119).

✔ **Chapter 6** includes a new short section on humans as weathering agents.

✔ **Chapter 7** contains a new section on subsidence and sedimentary basins.

✔ **Chapter 8** contains a new diagram of the relation of orogenic belts and plate boundaries and more material on fluids in metamorphism and evidence of veins.

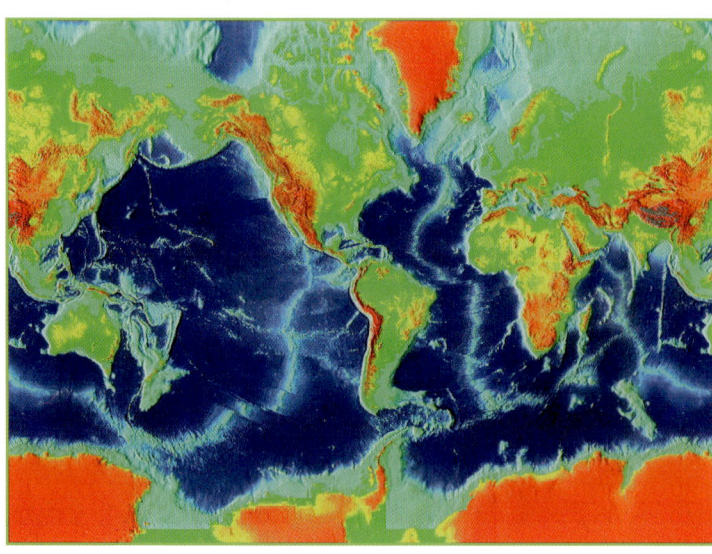

Figure 17.27 Topographic map for several figures, including 1.12 (p. 15), 18.14, 18.15, and 20.3. Provided by Peter W. Sloss, NOAA-NESDIS-NGDC, this shaded relief mercator projection shows both continental and seafloor topography and highlights crustal plate boundaries.

- ✔ **Chapter 9** includes new information on sequence stratigraphy, also describes how anthropologists, archaeologists, and seismologists use the geologic time scale.
- ✔ In **Chapter 12**, the authors have added an explanation of the effects of global warming on the hydrological cycle.
- ✔ **Chapter 13** now synthesizes the effect of tectonics and climate on deltas. The chapter also includes a new section on lakes and added information on the climate control of stream profiles and grades.
- ✔ **Chapter 14** now contains additional material on global wind belts.
- ✔ Embedded in **Chapter 15** are new theories on the origin of drumlins, and additional material on glaciation and climate, including expanded coverage of the Milankovitch cycles (with diagrams, ice borehole chronologies, isotopes and carbon dioxide in bubbles).
- ✔ **Chapter 18** now introduces the moment magnitude scale and discusses blind thrusts, an added seismic concern for Southern Californians.
- ✔ **Chapter 19** alludes to the newly discovered rough boundary between core and mantle and the latest findings from seismic tomography.
- ✔ **Chapter 21** reflects the latest thinking on how continents grow.
- ✔ **Chapter 22** now contains more material on the current debate about drilling in the Arctic National Wildlife Refuge, the concepts of sustainable development, and the relative cost of renewable energy sources.
- ✔ **Chapter 23** now explains that we still face the issue of sustainable development in a world of newly industrializing economies and uncontrolled population growth, despite the fact that the shortages forecast in the 1960s and '70s did not come to pass. A new box, "The Great Canadian Diamond Rush," describes how geologist Charles Fipke's scientific reasoning and detective work led to the discovery of North America's only commercial diamond deposit.
- ✔ ...and much more

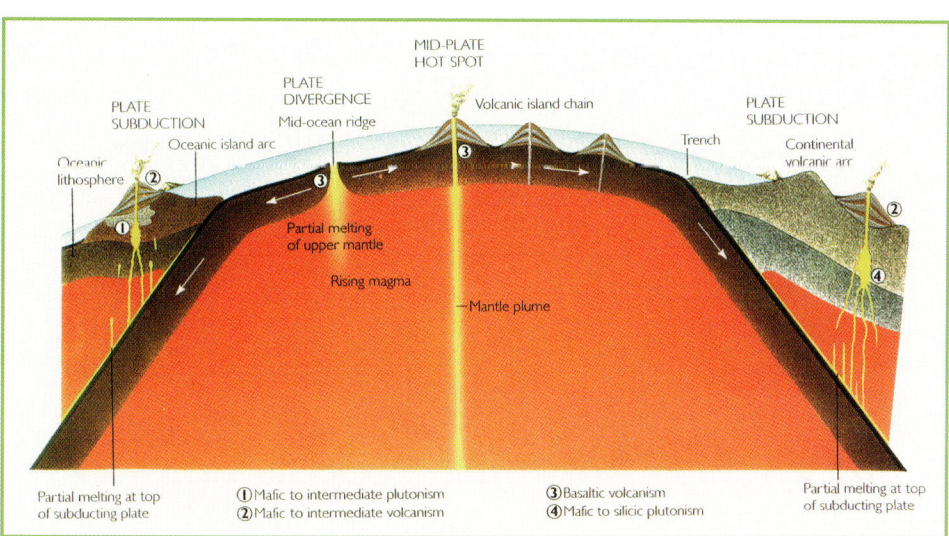

Figure 4.7 Plate tectonics and types of magma by Tomo Narashima

New Instructional Tools

BOXES

This edition contains an array of new boxed features, which highlight interesting topics or further explain difficult ones. Several existing features have been revised or matched with new illustrations.

To help students apply the boxed lessons to their growing knowledge of geology, the authors have classified them, using these beautiful color icons to represent the major themes of the book:

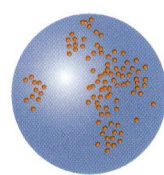

Living on Earth—These boxes either discuss ecologically sound behavior and wise use of the Earth's resources ("Water, A Precious Resource: Who Gets It" in Chapter 12) or describe how to prevent or mitigate disastrous geological events ("Protection in an Earthquake" in Chapter 18).

Interpreting the Earth— These boxes describe how geologists learn from the rocks around them.
- *New Box!* "Sedimentary Rocks Illuminate the Principle of Uniformitarianism" in Chapter 7.
- *New Box!* "Continental Drift: A Case History of the Scientific Method at Work" on pages 22 and 23 in Chapter 1 of this preview book.
- *New Box!* "Rifting Can Create a Sedimentary Basin" in Chapter 7.

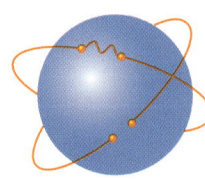

Technology and Earth—These boxes highlight the inventions and technology that help us study the Earth. For example, see "Kilauea: Monitoring a Volcano" on page 127 in Chapter 5 of your preview book.
- *Updated Box!* "Charting the Seafloor by Satellite" in Chapter 17.

To fully appreciate the instructive value of these boxes, please turn to pages 22 and 23 in Chapter 1 and peruse the authors' elegant explanation of the scientific method in relation to continental drift and the theory of plate tectonics.

24.1 INTERPRETING THE EARTH

Mass Extinctions and the Bolide - Impact Debate

About 65 million years ago, between the end of the Cretaceous period, and the beginning of the Tertiary, a mass extinction occurred in which half the life forms on Earth died out. Among the victims were the dinosaurs. Their disappearance allowed for the rapid evolutionary expansion in the Cenozoic era of mammals and, ultimately, humans.

Scientists have proposed several hypotheses to explain mass extinctions at the Cretaceous Tertiary (K-T) boundary. Most involve events or processes that could have induced widespread perturbations of Earth's systems. None has inspired more debate, more research, and ultimately more confidence than the bolide-impact hypothesis of K-T boundary extinctions proposed in 1980 by geologist Walter Alvarez. Alvarez and a team of scientists from the University of California at Berkeley were measuring the iridium content of sediment layers near Gubbio, Italy (see photo). Iridium is rare in Earth's crust and upper mantle but relatively abundant in the rest of the solar system. The iridium content of sedimentary layers, therefore, is a clue to the rates at which cosmic dust settled on Earth over time. If a slow but steady rain of cosmic dust has been settling on Earth, the iridium content of sedimentary layers deposited at various times should be similar.

Much to their surprise, the Alvarez team found concentrations of iridium in a 1-cm clay layer deposited exactly at the K-T boundary to be 30 times higher than in the sedimentary layers above and below it. Intrigued by this dramatic difference, the Alvarez team proposed that a large meteorite, perhaps 10 km in diameter, had collided with the earth at a speed of 75,000 km/hr or more. The force of impact

New Instructional Tools

NEW ILLUSTRATIONS
Close to 150 new illustrations, photos, maps

More than half the photographs and many of the maps and line drawings in this edition are new. The breathtaking shots illustrating and demonstrating the concepts students need to know include a pyroclastic eruption at Arenal Volcano in Costa Rica, crevasses in a New Zealand glacier, and a sulfur-encrusted fumerole in the Galápagos Islands. The array of gorgeous new maps includes the latest satellite altimetry map of the ocean floor and digital maps of crustal plates. This text is laden with portraits of the beauty, majesty, terrible destruction, and incomparable wealth that Earth's geologic forces can create.

Figure 16.2 Satellite photograph of Bathurst Island in Canada (© Earth Satellite Corporation)

TEAM PROJECTS

The Team Projects require students to explore geology through firsthand observation or research. Many of the projects underscore the need for cooperation among science, government, industry, and the public. Instructors can choose to assign these as short- or long-term take-home assignments. The projects appear at the end of selected chapters.

In addition to enhancing students' understanding of course material, Team Projects can strengthen students' researching, writing, and public speaking skills and encourage them to further explore the Internet. Jill S. Schneiderman, associate professor of geology at Vassar College, prepared the Team Projects.

Here are some examples:

✓ **Chapter 2 Short-Term Project: "Asbestos" (See page 56 of this preview book.)**
Students prepare a half-hour radio program on asbestos, assembling a list of experts with a variety of perspectives and presenting their arguments. The exercise shows students the value of a basic knowledge of mineralogy in deciding policy issues.

✓ **Chapter 3 Short-Term Project: "Identifying Building Stones" (See page 73.)**
Students examine the stone used in four to six buildings on campus or in their community. This project requires students to map the buildings, illustrate the stones, and draw a series of conclusions about such issues as the type of stone found, its origin, and whether the stone is well suited for its purpose. Students must explain and qualify their answers.

✔ **Chapter 9 Short-Term Team Project: "Geologic Time"**
This assignment asks students to develop their own metaphor for geologic time (for example: a life span, a mile, a metric ton) and then map a series of significant events, including the origin of the Earth, the first appearance of life on Earth, and students' own birthdays. The project is designed to give students a sense of the concept of geologic time.

✔ **Long-Term Project for Chapters 11, 12, 13, 14, 17, 18, 22, and 23**
Students monitor environmental changes—induced by nature or human interference—and local land-use policy decisions, and prepare monthly press releases, which could be sent to local media outlets. The press releases might address such issues as a local oil spill, the opening of a new landfill, or a flood. Students' releases should be brief, scientifically accurate, and written in terms the general public can understand.

INTERNET SOURCES

Internet addresses appearing at the end of each chapter help students and instructors tap into online information relevant to chapter material. Dallas Rhodes, Professor of Geology at Whittier College, compiled this lists of sites. Here are a few examples:

✔ Sites that feature images and text about the solar system augment the textbook material In Chapter 1, "Building a Planet."

✔ Sites addressing the subject of soils and acid rain enhance the material in Chapter 6, "Weathering and Erosion." These feature conference information, publications, and statistical data.

✔ Sites focusing on the Burgess Shale, the Grand Canyon, and a color-coded geologic timeline enhance the material in Chapter 9, "Geologic Time."

BUT WAIT—The Helpful Tools from the First Edition Are Still Here:

✔ **End of Chapter Summaries** review important concepts and help students assimilate new information.

✔ **Key Terms and Concepts Lists**, complete with page references, simplifies study and review of chapter material

✔ **Exercises and Thought Questions** at the end of each chapter test students' understanding of the material. Great for homework assignment or study material.

✔ **Suggested Readings** encourage students to augment their knowledge of the material and provide easy reference for instructors.

Figure 5.6 Pyroclastic eruption at Arenal Volcano in Costa Rica
(© Gregory G. Dimijian/Photo Researchers)

Frank Press

Frank Press has made pioneering contributions to the fields of geophysics, oceanography, lunar and planetary sciences, and natural resource exploration. His research has focused on the study of the Earth's crust and deep interior, and the seafloor. He built the instruments and was a member of the team that discovered the fundamental difference between oceanic and continental crust. In addition to his numerous distinctions and awards, Dr. Press holds the title of President Emeritus of the National Academy of Sciences in Washington, D.C. He is currently Cecil and Ida Green Senior Fellow at the Carnegie Institution of Washington's Department of Terrestrial Magnetism.

Dr. Press's eminence extends to the sphere of public policy, as well. He has advised four presidents on scientific issues. President Jimmy Carter appointed Dr. Press Science Advisor to the President and director of the Office of Science and Technology Policy. He served on presidential scientific advisory committees during the Kennedy and Ford administrations and as a member of the National Science Board under President Richard Nixon. Three times, *U.S. News and World Report* surveys named him the country's most influential scientist.

In the following interview, Dr. Press shares some of his thoughts on geology and on the second edition of *Understanding Earth*.

What do you enjoy most about geology?
I particularly value geology because it depends so much on observations of natural phenomena and draws upon many of the sciences to help understand what is observed in the field and laboratory.

How did you become interested in geology?
I was trained as a physicist and found that geology was an "outdoor" science with the world as my laboratory. I could travel, explore the oceans and continents, and do science at the same time.

What is your favorite aspect of the new edition of Understanding Earth?
Understanding Earth enables college students to understand how scientists work and think. At the same time they will obtain a solid foundation in a particular science—one that will enable them to appreciate the natural world around them. They will also see the relevance of science to public issues that will affect their lives profoundly. These are all related and represent what I like best about the book.

Raymond Siever

Raymond Siever is an internationally known expert in sedimentary petrology, geochemistry, and the evolution of oceans and the atmosphere. Dr. Siever is a longtime member of Harvard University's Department of Earth and Planetary Sciences, and he chaired the geology department for eight years. He was one of the first sedimentologists to apply the techniques of geochemistry to the study of sedimentary rock, especially sandstone and cherts.

In addition to co-writing the popular geology text, *Earth*, with Dr. Press, Dr. Siever co-wrote (along with F. J. Pettijohn and Paul Potter), the classic textbook *Sand and Sandstone* (Springer-Verlag). His book *Sand* (W. H. Freeman and Company) is a highly regarded addition to the *Scientific American Library* series. He is a Fellow of the Geological Society of America and the American Academy of Arts and Sciences and has been honored with distinguished awards from the Society of Sedimentary Geology, the Geochemical Society, and the American Association of Petroleum Geologists.

In this brief interview, Dr. Siever also shares some of his thoughts on his chosen field and the forthcoming edition of *Understanding Earth*.

What do you enjoy most about geology?
I most enjoy extracting basic ideas from the wonderful complexity of the natural world.

How did you become interested in geology?
When I was about eleven, my father packed us all into his new car and drove us out West. I saw the Grand Canyon, Zion and Bryce Canyons, and Monument Valley. I was goggle-eyed. All I did was look out the window at these rocks. I had never seen this before; I lived in the city—Chicago—surrounded by concrete.

In college, I planned to major in chemistry, but I wound up pursuing geology after taking a geology course just to see if I would like it. People often seemed surprised when I told them I was a geologist; it was a somewhat unusual career choice back then.

What is your favorite aspect of the new edition of Understanding Earth?
What Dr. Press and I like about *Understanding Earth* is very much the same thing we originally liked about *Earth*, our first textbook: We bring to elementary students some of the flavor of what is happening in science today. That attitude is what motivated us to write in the first place, and what continues to motivate us today.

The Supplements

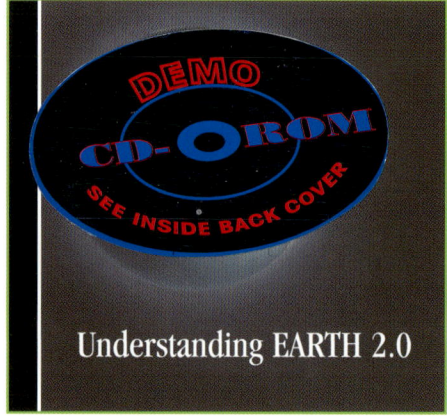

0-7167-2743-9

THE NEW EXPANDED CD-ROM FOR STUDENTS AND INSTRUCTORS ALIKE

An avalanche of videos, original animations, and photographs—many from sources outside the textbook—enliven the geological concepts described in the second edition of *Understanding Earth*. Other dramatic features of this innovative CD-ROM that benefit students and instructors include:

✔ **Q & A:** Students can use these electronic self-quizzes with built-in feedback to enhance their understanding of the material as well as to prepare for tests.

✔ **Illustrated Glossary:** The definitions and illustrations in this electronic glossary will help students remember the myriad of new terms they encounter.

✔ **Geographic Locator:** This study tool is designed to help students learn the relative locations of the hundreds of places referred to in the book.

✔ **Presentation Manager:** Instructors can easily queue up a sequence of images, videos, and original animations for display during lectures using this convenient tool.

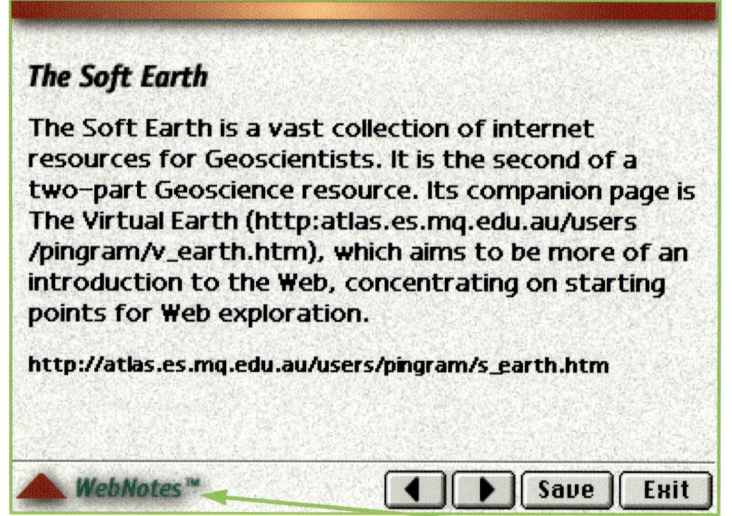

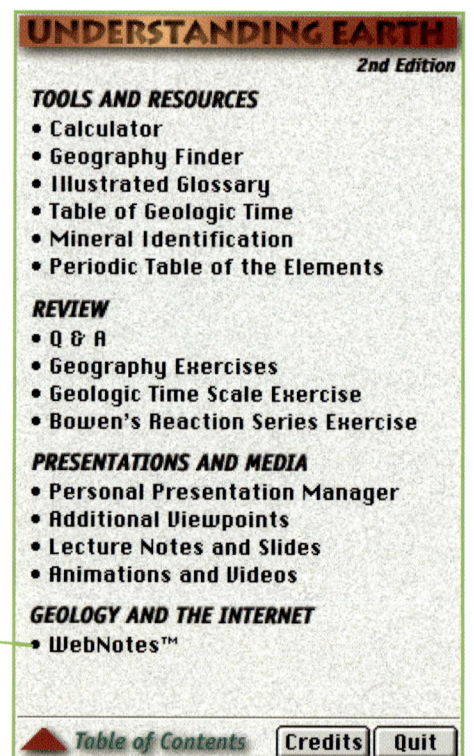

✔ **WebNotes™:** Students will find hundreds of geological URLs that refer to specific sections in the book.

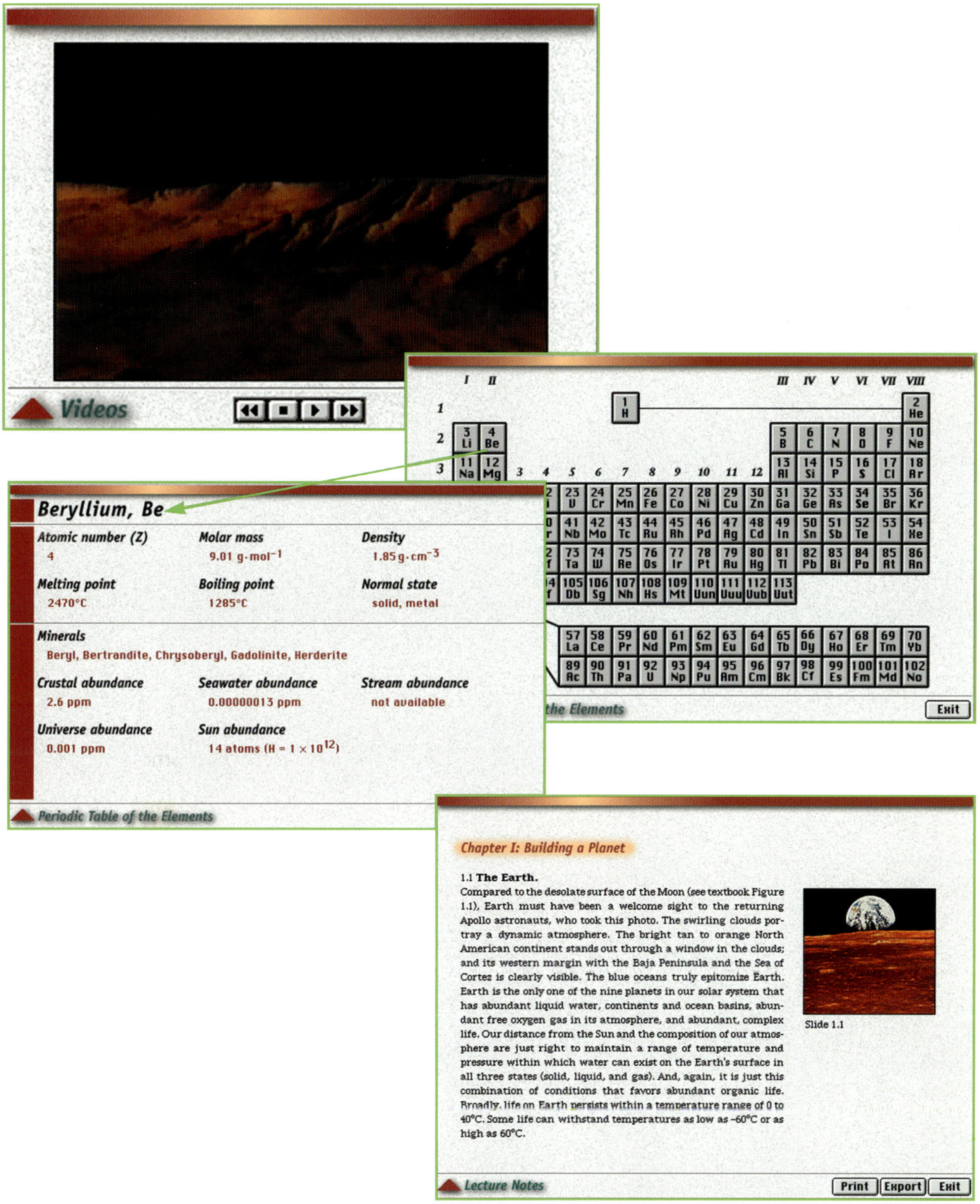

Press/Siever - UNDERSTANDING EARTH, 2/e

THE SUPPLEMENTS

✓ **Periodic Table:** The electronic periodic table provides such information as melting and boiling points of elements, their densities, and their abundance on Earth and elsewhere in the universe.

✓ **Slide Set with Lecture Notes:** The electronic version of Peter Kresan's slide/lecture note package for *Understanding Earth*, 2/e. (see p.16 of this overview)

The Supplements

TWO ENTIRELY NEW FEATURES

Press Releases and our **Understanding Earth Web Site** work together to help you demonstrate to students geology's relevance and importance. When an event of geological significance occurs, you will receive a **Press Release** containing support materials that will help you convey it vividly to your students.

The **Understanding Earth Web Site** provides up-to-date information, new images every month that can be exported into presentation programs, as well as offering updated WebNotes™ and self-quizzes for your students. A preview of our Web site will be posted on January 1, 1997. The address is www.whfreeman.com/understandingearth

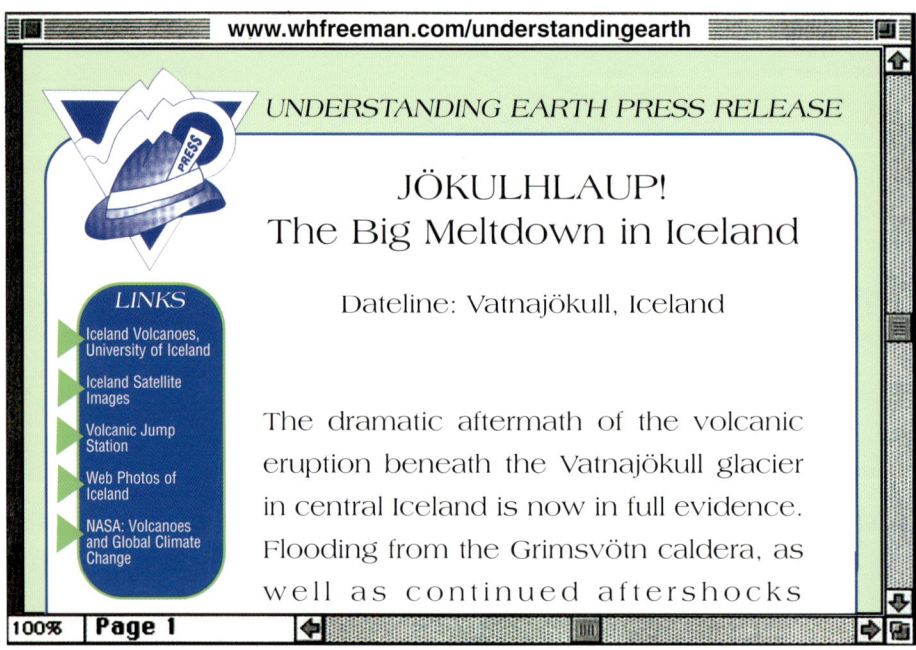

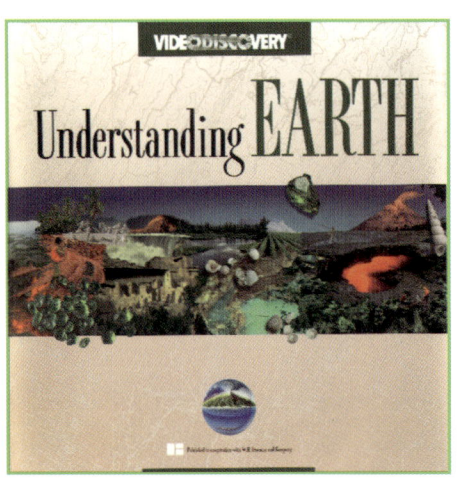

THE UNDERSTANDING EARTH VIDEODISC

produced by Videodiscovery in cooperation with W. H. Freeman and Company. Still the best source of images on videodisc for the geology classroom.

0-7167-2590-8

Press/Siever - UNDERSTANDING EARTH, 2/e

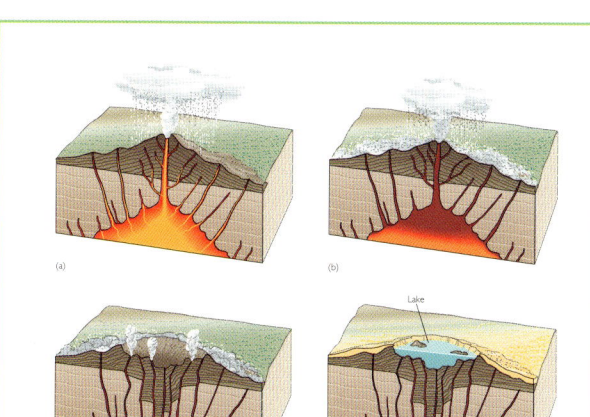

Figure 5.16 The evolution of a caldera by Ian Worpole

OVERHEAD TRANSPARENCY SET
0-7167-2888-5

This set contains 110 dazzling color diagrams reproduced from the text. Some transparencies come with overlays and labels, which, where appropriate, contain oversized type.

SLIDE SET WITH LECTURE NOTES
0-7167-2813-3

Peter L. Kresan at the University of Arizona has expanded this popular slide set to include 100 more images, bringing the total to 220 images not found in the text. These slides will enhance and illustrate your lectures. As with the earlier set, each slide is accompanied by a thumbnail representation and thorough lecture notes describing the image's setting and geological importance.

Taku and Norris glaciers, Juneau Icefield, Alaska © Peter Kresan

---

Reserve Your Copy of the
NEW EDITION
of
UNDERSTANDING EARTH
T O D A Y !

For a complimentary review copy of Press and Siever's UNDERSTANDING EARTH, Second Edition, please complete and return this postpaid card. You may also call toll-free

1-800-446-8923

Yes! Please send me Press and Siever's **UNDERSTANDING EARTH**, 2/e, 0-7167-2836-2

Name:
Department:
School: <place address label here>
Address:
City/State/Zip:
E-mail:
Course name and number:
Course length: ○ One semester ○ Two semesters ○ _____ Quarters
Annual enrollment: Decision date:
Current text in use:(author/title)

What do you like/dislike about your current book?
Likelihood of change: ○ High ○ Medium ○ Low
Office Hours: Phone: Best time to call:
___ ___ ___ 6910
ST   SCH REP JOB

The Supplements

THOROUGHLY REVISED TEST BANK- Printed 0-7167-2795-1; MAC: 0-7167-2780-3; Windows: 0-7167-2750-1

Many questions now contain geological diagrams. The computerized versions—Windows and Macintosh—allow instructors to easily change and add questions as well as to import their own electronic drawings. The revised *Test Bank* was written by Simon M. Peacock of Arizona State University.

THOROUGHLY UPDATED STUDY GUIDE- 0-7167-2804-4

Students will find a vast array of exercises in this guide to help them solidify their grasp of the key concepts in *Understanding Earth*. For instructors, the guide can serve as a source of ready-made homework assignments. David M. Best of Northern Arizona University, who prepared the guide, has updated the material for this edition. A new addition, the "CD Lab," helps students get the most from the CD-ROM by calling upon them to interpret specific content and answer questions.

INSTRUCTOR'S RESOURCE MANUAL- 0-7167-2765-X

In addition to the more traditional features of a resource manual, instructors will find information on using the CD-ROM, the Web site, and other items in the supplements package for *Understanding Earth*. The manual also suggests ways to use the Web and other electronic media in the geology classroom and lab. The *IRM* was written by Philip M. Astwood of the University of South Carolina, Columbia.

BUSINESS REPLY MAIL
FIRST CLASS MAIL PERMIT NO. 7953 NEW YORK, NY

Postage Will Be Paid By Addressee

W. H. FREEMAN AND COMPANY

c/o Scientific American/St. Martin's Press College Desk
345 Park Avenue South
New York, NY 10160-1039

NO POSTAGE
NECESSARY
IF MAILED
IN THE
UNITED STATES

Understanding Earth

Understanding Earth
SECOND EDITION

FRANK PRESS
Carnegie Institution of Washington

RAYMOND SIEVER
Harvard University

W. H. Freeman and Company
NEW YORK

TO OUR CHILDREN, AND OUR CHILDREN'S CHILDREN;
MAY THEY LIVE IN HARMONY WITH EARTH'S ENVIRONMENT.

ACQUISITIONS EDITOR: *Holly Hodder*
DEVELOPMENT EDITORS: *Susan Seuling; Nancy Fleming*
PROJECT EDITORS: *Mary Louise Byrd; Penelope Hull*
TEXT AND COVER DESIGNER: *Vertigo Design*
COVER ILLUSTRATION: *Tomo Narashima*
INTERIOR ILLUSTRATIONS: *Ian Warpole, Network Graphics, and Tomo Narashima*
PHOTO RESEARCH: *Alexandra Truitt; Jerry Marshall*
PRODUCTION COORDINATOR: *Paul W. Rohloff*
COMPOSITION: *York Graphic Services*
MANUFACTURING: *RR Donnelley & Sons Company*

LIBRARY OF CONGRESS CATALOGING-IN-PUBLICATION DATA

Press, Frank.
Understanding earth / Frank Press, Raymond Siever.—2nd ed.
p. cm.
Includes index
ISBN 0-7167-2836-2   ISBN 0-7167-3099-5 (minibook)
1. Earth sciences.   I. Siever, Raymond.   II. Title.
QE28.P9   1997
550—dc21   96-43804
CIP

© 1998, 1994 by W. H. Freeman and Company

No part of this book may be reproduced by any mechanical, photographic,
or electronic process, or in the form of any phonographic recording, nor may
it be stored in a retrieval system, transmitted, or otherwise copied for public or
private use, without written permission from the publisher.

Printed in the United States of America.
First printing, 1997

Brief Contents

Part 1   Understanding the Earth System   1

1. Building a Planet   2
2. Minerals: Building Blocks of Rocks   26
3. Rocks: Records of Geologic Processes   58
4. Igneous Rocks: Solids from Melts   74
5. Volcanism   104
6. Weathering and Erosion
7. Sediments and Sedimentary Rocks
8. Metamorphic Rocks
9. The Rock Record and the Geologic Time Scale
10. Folds, Faults, and Other Records of Rock Deformation

Part 2   Surface Processes

11. Mass Wasting
12. The Hydrologic Cycle and Groundwater
13. Rivers: Transport to the Oceans
14. Winds and Deserts
15. Glaciers: The Work of Ice
16. Landscape Evolution
17. The Oceans

Part 3   Internal Processes, External Effects

18. Earthquakes
19. Exploring Earth's Interior
20. Plate Tectonics: The Unifying Theory
21. Deformation of the Continental Crust

Part 4   Conserving Earth's Bounty

22. Energy Resources from the Earth
23. Mineral Resources from the Earth
24. Earth Systems and Cycles

Contents

TO THE INSTRUCTOR   xi
TO THE STUDENT   xvii

Part 1 Understanding the Earth System   1

1 Building a Planet   2
The Scientific Method   4
The Principle of Uniformitarianism   4
The Origin of Our System of Planets   4
Earth as an Evolving Planet   8
Plate Tectonics: A Unifying Theory for Geological Science   14
Geologists at Work   20

2 Minerals: Building Blocks of Rocks   26
What Are Minerals?   28
The Atomic Structure of Matter   29
Chemical Reactions   30
Chemical Bonds   35
Atomic Structure of Minerals   35
Rock-Forming Minerals   40
Physical Properties of Minerals   45

3 Rocks: Records of Geologic Processes   58
Igneous Rocks   61
Sedimentary Rocks   62
Metamorphic Rocks   64
The Chemical Composition of Rocks   65
Where We See Rocks   65
The Rock Cycle   68
Plate Tectonics and the Rock Cycle   70

4 Igneous Rocks: Solids from Melts   74
How Do Igneous Rocks Differ From One Another?   76
How Do Magmas Form?   83
Where Does Magma Form?   85
Magmatic Differentiation   86
Forms of Magmatic Intrusions   94
Igneous Activity and Plate Tectonics   98

5 Volcanism   104
Volcanic Deposits   107
Eruptive Styles and Landforms   111
The Global Pattern of Volcanism   121
Volcanism and Human Affairs   125

6 Weathering and Erosion
Weathering, Erosion, and the Rock Cycle
Why Do Some Rocks Weather More Rapidly Than Others?
Chemical Weathering
Physical Weathering
Soil: The Residue of Weathering
Humans as Weathering Agents
Weathering Makes the Raw Material of Sediment

7 Sediments and Sedimentary Rocks
Sedimentary Rocks and the Rock Cycle
Sedimentary Environments
Sedimentary Structures

Contents

Burial and Diagenesis: From Sediment to Rock
Classification of Clastic Sediments and Sedimentary Rocks
Classification of Chemical and Biochemical Sediments and Sedimentary Rocks

8 Metamorphic Rocks

Causes of Metamorphism
Physical and Chemical Forces Controlling Metamorphism
Kinds of Metamorphism
Metamorphic Textures
Regional Metamorphism and Metamorphic Grade
Contact Metamorphic Zones
Plate Tectonics and Metamorphism

9 The Rock Record and the Geologic Time Scale

Timing the Earth
Reconstructing Geologic History Through Relative Dating
Radiometric Time: Adding Dates to the Time Scale
From Three Lines of Evidence: A Reliable Dating Tool

10 Folds, Faults, and Other Records of Rock Deformation

Interpreting Field Data
How Rocks Become Deformed
How Rocks Fold
How Rocks Fracture: Joints and Faults
Unraveling Geologic History

Part 2 Surface Processes

11 Mass Wasting

What Makes Masses Move?
Classification of Mass Movements
Understanding the Origins of Mass Movements

12 The Hydrologic Cycle and Groundwater

Flows and Reservoirs
Hydrology and Climate
The Hydrology of Runoff
Groundwater
Water Resources from Major Aquifers
Erosion by Groundwater
Water Quality
Water Deep in the Crust

13 Rivers: Transport to the Oceans

How Stream Waters Flow
Stream Loads and Sediment Movement
How Running Water Erodes Solid Rock
Stream Valleys, Channels, and Floodplains
Streams Change with Time and Distance
Drainage Networks
Deltas: The Mouths of Rivers

Contents

14 Winds and Deserts
- Wind as a Flow of Air
- Wind as a Transport Agent
- Wind as an Agent of Erosion
- Wind as a Depositional Agent
- The Desert Environment

15 Glaciers: The Work of Ice
- Ice as a Rock
- What Is a Glacier?
- Glacial Budgets: How Glaciers Form, Grow, and Shrink
- How Glaciers Move
- Glacial Landscapes
- Ice Ages: The Pleistocene Glaciation

16 Landscape Evolution
- Topography, Elevation, and Relief
- Landforms: The Components of Landscape
- Factors That Control Landscape
- The Face of North America
- The Evolution of Landscape

17 The Oceans
- The Edge of the Sea: Waves and Tides
- Shorelines
- Sensing the Floor of the Ocean
- Profiles of Two Oceans
- Continental Margins
- The Floor of the Deep Ocean
- Sedimentation in the Sea
- Differences in the Geology of Oceans and Continents

Part 3 Internal Processes, External Effects

18 Earthquakes
- What Is an Earthquake?
- Studying Earthquakes
- The Big Picture: Earthquakes and Plate Tectonics
- Earthquake Destructiveness

19 Exploring Earth's Interior
- Exploring the Interior with Seismic Waves
- Earth's Internal Heat
- The Interior Revealed by Earth's Magnetic Field

20 Plate Tectonics: The Unifying Theory
- From Controversial Hypothesis to Respectable Theory
- Overview
- The Mosaic of Plates
- Rates of Plate Motion
- The Geometry of Plate Motion
- Rock Assemblages and Plate Tectonics

Contents

    Microplate Terranes
      and Plate Tectonics
    The Grand Reconstruction
    The Driving Mechanism
      of Plate Tectonics

21 Deformation of the Continental Crust

    Some Regional Tectonic Structures
    The Stable Interior
    Orogenic Belts
    Coastal Plain and Continental Shelf
    Regional Vertical Movement

Part 4 Conserving Earth's Bounty

22 Energy Resources from the Earth

    Resources and Reserves
    Energy
    Oil and Natural Gas
    Coal
    Oil Shale and Tar Sands
    Alternatives to Fossil Fuels
    Nuclear Energy Fueled by Uranium
    Solar Energy
    Geothermal Energy
    Conservation
    Energy Policy

23 Mineral Resources from the Earth

    Minerals as Economic Resources
    The Geology of Mineral Deposits
    Ore Deposits and Plate Tectonics
    Finding New Mineral Deposits

24 Earth Systems and Cycles

    Formation of Earth Systems
    Geochemical Cycles: Tracers
      of Earth's Systems
    Geological Perturbances
      and Climate Change
    Human Activity and
      Global Change

Appendix 1

    Conversion Factors

Appendix 2

    Numerical Data Pertaining to Earth

Appendix 3

    Properties of the Most Common
      Minerals of Earth's Crust

Appendix 4

    Topographic and Geologic Maps

Glossary

Index

Boxed Features

 ### Living on Earth

- 2.1 What Makes Gems So Special?
- 2.2 Asbestos and Health
- 6.1 Soil Erosion
- 6.2 Mapping Soil Types with a Newer Classification
- 9.2 Radon: An Environmental Threat
- 11.1 Preventing Landslides
- 11.2 Reducing Loss from Landslides
- 12.1 Water, A Precious Resource: Who Gets It?
- 12.2 When Do Groundwaters Become Nonrenewable Resources?
- 13.1 The Development of Cities on Floodplains
- 13.2 Historic Floods and Flood Control
- 14.1 Droughts and Dust Bowls
- 17.1 Preserving Our Beaches
- 17.3 The Oceans as a Deep-Waste Repository
- 18.1 Tsunamis
- 18.2 Protection in an Earthquake
- 22.1 Radioactive Waste Disposal
- 23.1 Use of Federal Lands in the United States

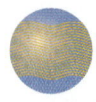 ### Interpreting the Earth

- 1.1 Continental Drift: A Case History of the Scientific Method at Work
- 4.1 Japan: A Growing Island Arc
- 5.1 The Explosion of Krakatoa
- 5.3 Mount St. Helens: Dangerous but Predictable
- 7.1 Sedimentary Rocks Illuminate the Principle of Uniformitarianism
- 7.2 Rifting Can Create a Sedimentary Basin
- 8.1 New England Metamorphic Terrane
- 9.1 The Grand Canyon Sequence
- 14.2 Desertification on the Fringe of the Sahara
- 15.1 Future Changes in Sea Level and the Next Glaciation
- 17.2 Hot Springs on the Seafloor
- 19.2 The Uplift of Scandinavia: Nature's Experiment with Isostasy
- 21.1 The Collision between India and Eurasia

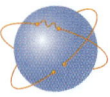 ### Technology and Earth

- 5.2 Kilauea: Monitoring a Volcano
- 17.3 Charting the Seafloor by Satellite
- 19.1 Finding Oil with Seismic Waves
- 20.1 Drilling in the Deep Sea
- 24.1 Mass Extinctions and the Bolide-Impact Debate

To the Instructor

When we wrote our first textbook, *Earth,* geology was a field flush with the excitement of new discovery. Recognition of continental drift only a decade earlier had triggered a revolution in our understanding of our planet. For the first time in the history of the discipline, an all-encompassing synthesis of geological knowledge was being advanced. Plate-tectonic theory gave us a framework for learning about the immense forces turning cyclically in the core, in the surrounding mantle, in the crust, and in the air, oceans, and biosphere, keeping our planet in a constant state of change. This new picture of Earth as a dynamic, coherent system was central to our writing of that book and its successor, *Understanding Earth.* We wanted to share with as many students as possible something of the exhilaration and intellectual excitement the profession was feeling. We have been enormously gratified by the warmth and loyalty with which each edition of *Earth,* and later *Understanding Earth,* was received.

More than thirty years into the plate-tectonics revolution, an understanding of the whole Earth system is more necessary than ever. The world in which we live has changed. With its growing population and increasing industrial societies, our world is using more resources, contributing more to pollution of air, water, and land, and becoming more vulnerable to environmental disaster than ever before. Human disturbances of the environment now equal or exceed natural disturbances in magnitude and rate. Our planet enjoys a unique and delicate balance of conditions that allowed life to begin and countless life forms to flourish. Because they affect multiple systems and ultimately the whole Earth, relentless abuses of the environment are a threat to life on the planet.

Students now entering college belong to the generation that will lead our world through the first decades of the next century. We believe the social, political, and economic issues they face will prove many times more challenging than those we have already encountered. To make wise decisions about such issues as resource development, waste management, environmental protection, and land use, they will have a tremendous need for scientific literacy in general and an understanding of geology in particular. We have brought this conviction to our textbook.

Understanding Earth is for today's students, especially beginning students whose one course in geology may be their sole college exposure to the physical sciences. Naturally, as geologists we want to share many of the fascinating aspects of our discipline with students. As teachers we have tried to do this in a way that is compelling, accurate, and above all, up to date. We do so with the hope that learning how scientists think and work, and understanding something of the geological systems and processes that underlie the past, present, and future of our planet, will help our readers to think more deeply and responsibly about the issues they will confront as citizens.

GOALS

Understanding Earth is designed for a one-term introductory course in physical geology for non-science majors. We took it as a challenge to present the essential material, both traditional and modern, that a good geology course should cover, but in terms accessible to a student who has had no previous college science. We have deliberately emphasized a broad view, one stressing concepts and trying to show by many examples what is meant by the "scientific method." We have tried to impart something of what motivates contemporary geologists. As much as possible, we have described contemporary technology and methods and discovery. We have tried to integrate in a natural way the newest discoveries and environmental concerns with the traditional discussions of such basic topics as geomorphology, sedimentation, petrology, volcanism, and structural geology.

PEDAGOGY

- **Illustrations** The explanatory drawings, diagrams, and photos in *Understanding Earth* are of considerable value in simplifying otherwise difficult concepts. These illustrations and their captions serve as alternative restatements of concepts presented in the text and as intermittent summaries. Much of our attention in this edition was devoted to making the proven illustrations even clearer and adding more color photos. We replaced more than half the photos in this edition and included an even broader range of photographic examples from all over the world.

- **Maps** Maps are essential to understanding plate tectonics and consequent phenomena such as earthquakes and volcanoes. In addition to the exceptionally clear schematic world maps in the first edition, the second edition of *Understanding Earth* incorporates global relief maps designed by Dr. Peter W. Sloss under the auspices of the National Geophysical Data Center.
- **Boxed Features** The program of boxed features has been expanded and refined to reflect the major themes of the book: Living on Earth, Interpreting the Earth, and Technology and Earth.

    Living on Earth boxed features carry forward the theme of environmental awareness by tackling such issues as mitigating natural catastrophes and using resources wisely in ways that preserve the Earth.

    Interpreting the Earth boxed features carry forward the theme of scientific discovery made through observation and logic. This category includes features on scientific method and the meaning of particular geologic formations.

    Technology and Earth boxed features look at how geologists use technology to gather information. This category includes features on seafloor and ice core drilling, monitoring volcanoes, and satellite altimetry mapping of the seafloor.

- **Study and Research Aids** Each chapter begins with an engaging story or piece of information followed by a statement of goals that offers a unifying view of the chapter in a manner that is conceptually accessible to the beginning student. **Key terms** are highlighted in bold type within each chapter and are listed at the end of the chapter for easy reference and review. Key terms and definitions are compiled in the **Glossary** at the back of the book. **Chapter Summaries** serve as systematic reviews of major concepts. **Exercises** help students test their comprehension of key chapter material. **Thought Questions** ask students to apply ideas and principles to situations not specifically covered in the text. **Suggested Readings** point curious students toward books and articles, both popular and technical, on subjects they wish to explore more deeply. Internet sources direct students to the latest research and a variety of related materials available on Web sites. **Team Projects,** at the end of selected chapters, offer instructors a variety of inventive ideas for short-term, long-term, written, and oral group assignments.

ORGANIZATION

To accommodate the many ways that instructors may want to structure the course, we have made each chapter as self-sufficient as possible. Nevertheless, few geologic processes can be taught as wholly independent subjects; they must be seen in the context of a larger system. We have thus used the recurrence of many important topics as an opportunity for review and alternative restatement. This approach enhances learning and increases flexibility in the way the book can be used.

Relying on a consensus of views from colleagues and reviewers, we arrived at the following four-part organization for this text.

PART 1 UNDERSTANDING THE EARTH SYSTEM   Part 1 could almost be considered a basic text in the foundations of geology that evolved in the first half of the century, illuminated by the insights of plate-tectonic theory. An introductory chapter proceeds from a discussion of the origin of Earth to a first treatment of the geological cycle and elementary plate tectonics. This enables the instructor to relate plate tectonics to the many subjects covered before Chapter 20, the chapter devoted solely to plate tectonics.

A comprehensive chapter on mineralogy is followed by a short chapter introducing the rock cycle and the three major classes of rocks. The advantage of the rock cycle chapter is that it establishes the relationship of the rock cycle to other Earth systems and offers the instructor flexibility in assigning the more intensive chapters on each class of rock. The chapter on igneous rocks is followed by the closely related chapter on volcanism, so that knowledge of the origin of igneous rocks precedes the particulars of eruptive processes. In contrast, the weathering and erosion chapter is a logical prelude to the origin of sedimentary rocks in the following chapter. Instructors who choose to reverse the order of presentation of igneous and sedimentary rocks will find that the text is adaptable and flexible in this regard. After the major rock types are covered, the student is introduced to geologic time, stratigraphy, and structure.

PART 2 SURFACE PROCESSES   Part 2 covers the major topics of geomorphology: water, rivers, wind, and ice. These subjects are the most easily comprehended areas of geology because most students can draw on personal experience with Earth's landscapes. Completing this part is a chapter on the oceans, which continue to grow in importance in geology and in global environmental issues. Here we emphasize shorelines and shallow water processes, because they are the most readily observed, and we discuss those aspects of the deep sea that relate most closely to plate tectonics and to the formation of marine sediments. Part 2 examines the surface systems individually and in relation to each other and to the interior.

PART 3 INTERNAL PROCESSES, EXTERNAL EFFECTS Part 3 is an exploration of Earth's interior and its dynamic interaction with the crust. A chapter on earthquakes leads into a chapter on what seismic waves, heat flow, and magnetism reveal about the structure and behavior of Earth's interior. This chapter leads into a full treatment of plate tectonics as it is understood today. Following is a chapter on the role of plate tectonics in the formation and deformation of the continents, a culmination that brings together all the elements of geologic history, which defied interpretation in the earlier decades of this century.

PART 4 CONSERVING EARTH'S BOUNTY Part 4 begins with chapters on mineral and energy resources. These chapters weigh economic benefit against environmental degradation and highlight the advantages of policies that encourage economic development and use of nonpolluting energy sources such as solar and wind power.

The last chapter, "Earth Systems and Cycles," is new to this edition. This chapter develops the theme of interdependency among geosphere, atmosphere, hydrosphere, and biosphere, and ends with an eloquent argument for a responsible approach to using Earth's resources.

Trends and Themes

Plate tectonics continues to inspire a fast pace of activity in today's geology, geochemistry, and geophysics. To lend currency and excitement to the introductory course, we integrate, at an appropriate level, the flavor of today's research with the basic principles of physical geology.

We have introduced a number of topics that are the focus of current research in important areas of geology. Among these topics are the relationship of sedimentary sequences to sea level, the evolution of orogenic belts in relation to plate tectonics, current ideas on glaciation and climate, crust–mantle interactions, and the influence of comet and asteroid impacts on Earth's surface environments.

We call attention to newer approaches to some subjects, for example, correlating sea-level changes with climate and tectonics and interpreting continental geology in terms of microplates and ancient plate interactions.

We introduce students to newer technologies, such as side-scan acoustic radar used in oceanography and seismic wave tomography used in the study of the interior. We want to impart the idea that geology, like other sciences, is constantly changing, renewing itself as new ideas and technology come to the fore.

Geology has been and remains an eminently practical science. We discuss geologic hazards and hazard prevention in chapters on volcanism (dangerous eruptive behavior), weathering and erosion (soil loss), mass wasting (landslides), rivers (floods), water (groundwater contamination), and the interior (earthquakes). We also discuss the economic and social issues that hinge on geology.

All the common metals used in industry originate as ore deposits discovered by geologists. Sand, gravel, and limestone are used for construction and for many other industrial activities. Our major sources of energy—coal, oil, and gas—are geological deposits. Exploration for oil and gas, in fact, remains one of the major occupations of geologists. We discuss reasons for extracting and using these resources in ways that minimize damage to the Earth, and we look at the benefits and costs of alternative energy sources.

Global change is seen increasingly as a threat to life on Earth. We cover the connections among climate and weathering and erosion, landscape, and sedimentation throughout the book and devote the new, final chapter to the interaction of Earth's systems. We discuss climate in relation to natural and human-induced conditions that trigger changes in glaciation, the greenhouse effect, and the amount of carbon dioxide in the atmosphere.

One of the most important aspects of geology is its role in the evolution of life on Earth. Geology is a framework for understanding the physical conditions—surface changes and climate changes—underlying the myriad forms of life that evolved over billions of years. Global climate change is seen increasingly as a threat to life.

SUPPLEMENTS

We are again proud to announce the availability of a vast selection of print, visual, and electronic supplements for students and instructors using *Understanding Earth*. Many are completely new to this edition and take advantage of the best and latest technology.

- ***CD-ROM*** An innovative *CD-ROM* has been developed especially for this book to bring an added dimension to *Understanding Earth*. Among the features are **Q&A**—electronic self-quizzes with built-in feedback that students can use to practice for tests; **Added Views**—a wide selection of geological videos and original animations, as well as photographs that complement images found in the book and expanded topic coverage; **Illustrated Glossary**—an electronic device that defines, illustrates, and helps students recall the hundreds of terms used in introductory geology; **WebNotes™**—hundreds of geological URLs referenced to specific sections in the book; **Geographic Locator**—an electronic tool designed to show and help students learn the relative locations of map sites mentioned in the book; **Presentation Manager**—enables instructors to easily queue up a sequence of images, videos, and animations for display during lectures; **Tools and Resources**—a collection of components such as unit converter, calculator, interactive periodic table, and interactive geologic time scale to help the students interact with various geologic concepts.

- ***Study Guide*** David M. Best of Northern Arizona University has thoroughly updated his *Study Guide to Understanding Earth*. Many exercises use illustrations from the book and all were designed to focus on the key content in each chapter. Special "CD Lab" exercises help students see the most effective use of the CD-ROM. The *Study Guide* also contains detailed chapter summaries, complete chapter outlines, and practice multiple-choice tests.

- ***Instructor's Resource Manual*** Tying together the selection of supplements for instructors is the *Instructor's Resource Manual*, by Philip M. Astwood of the University of South Carolina, Columbia. In addition to the more traditional features, the *Manual* contains information on using the CD-ROM, the Web Site, and other items in the supplements package. It also offers ideas on using some of the newer electronic products available in the geology classroom and lab.

- ***Slide Set with Lecture Notes*** Peter L. Kresan of the University of Arizona has augmented the first-edition *Slide Set* with an additional 100 images and lecture notes carefully selected to enhance the images in the book. This brings the total number of slides offered to 220.

- ***Overhead Transparency Set*** About 110 vivid color diagrams are reproduced from the text in the *Overhead Transparency Set*. Some come with overlays of the labels. Oversized type is used for labels where appropriate.

- ***Test Bank*** Simon M. Peacock of Arizona State University has thoroughly revised the *Test Bank* for the new edition. A sizable number of the questions now contain geological diagrams, and the Windows and Macintosh software of the computerized versions allow instructors easily to change and add questions as well as to import their own electronic drawings.

- ***Understanding Earth Videodisc*** Produced by Videodiscovery, Inc., in cooperation with W. H. Freeman and Company, the *Understanding Earth Videodisc* is again available to qualified adopters.

- ***Press Releases*** This new supplement provides instructors with support materials on geologically significant events on an ongoing basis.

- ***Understanding Earth Web Site*** Also new to this edition, this electronic supplement provides continual support to the book and the CD-ROM. In addition to the information about important geological events, the site provides new images every month that can be exported into presentation programs, updated WebNotes™, another source of self-quizzes for students, and other features. Its address is: www.whfreeman.com/understandingearth

ACKNOWLEDGMENTS

It is a challenge both to teachers and to authors of geology texts to encompass the many important aspects of geology in a single course and to inspire interest and enthusiasm in the student. To meet this challenge, we have called on the advice of many colleagues teaching in all kinds of college and university settings. From the earliest planning stages of each edition of this book we relied on a consensus of views in deciding on an organization for the text and in choosing which topics to include. As we wrote and rewrote the chapters, we again relied on our colleagues to guide us in making the presentation pedagogically sound, accurate, and accessible and stimulating to students. To each one we are grateful.

Wayne M. Ahr
Texas A & M
Gary Allen
University of New Orleans
N. L. Archbold
Western Illinois University
Allen W. Archer
Kansas State University
Richard J. Arculus
University of Michigan, Ann Arbor
Philip M. Astwood
University of South Carolina
R. Scott Babcock
Western Washington University
Evelyn J. Baldwin
El Camino Community College
Lukas P. Baumgartner
University of Wisconsin-Madison
David M. Best
Northern Arizona University
Dennis K. Bird
Stanford University
Stuart Birnbaum
University of Texas, San Antonio
David L. S. Blackwell
University of Oregon
Arthur L. Bloom
Cornell University
Phillip D. Boger
State University of New York, Geneseo
Robert L. Brenner
University of Iowa
David S. Brumbaugh
Northern Arizona University
Edward Buchwald
Carleton College
Robert Burger
Smith College
Timothy Byrne
University of Connecticut
J. Allan Cain
University of Rhode Island
F. W. Cambray
Michigan State University
Ernest H. Carlson
Kent State University
Max F. Carman
University of Houston
Allen Cichanski
Eastern Michigan University
George R. Clark II
Kansas State University
G. S. Clark
University of Manitoba

Roger W. Cooper
Lamar University
Peter Dahl
Kent State University
Jon Davidson
University of California, Los Angeles
Larry E. Davis
Washington State University
Robert T. Dodd
State University of New York at Stony Brook
Bruce J. Douglas
Indiana University
Grenville Draper
Florida International University
William M. Dunne
University of Tennessee, Knoxville
R. Lawrence Edwards
University of Minnesota
C. Patrick Ervin
Northern Illinois University
Stanley Fagerlin
Southwest Missouri State University
Jack D. Farmer
University of California, Los Angeles
Stanley C. Finney
California State University, Long Beach
Tim Flood
Saint Norbert College
Richard M. Fluegeman, Jr.
Ball State University
Michael F. Follo
Colby College
Richard L. Ford
Weaver State University
Charles Frank
Southern Illinois University
William J. Frazier
Columbus College
Robert B. Furlong
Wayne State University
Sharon L. Gabel
State University of New York, Oswego
Alexander E. Gates
Rutgers University
Gary H. Girty
San Diego State University
William D. Gosnold
University of North Dakota
Richard H. Grant
University of New Brunswick
Bryan Gregor
Wright State University

G. C. Grender
Virginia Polytechnic Institute and State University
Mickey E. Gunter
University of Idaho
David A. Gust
University of New Hampshire
Kermit M. Gustafson
Fresno City College
Bryce M. Hand
Syracuse University
Ronald A. Harris
West Virginia University
Eric Hetherington
University of Minnesota
J. Hatten Howard III
University of Georgia
Herbert J. Hudgens
Tarrant County Junior College
Ian Hutcheon
University of Calgary
Ruth Kalamarides
Northern Illinois University
Frank R. Karner
University of North Dakota
Phillip Kehler
University of Arkansas, Little Rock
Peter L. Kresan
University of Arizona
Albert M. Kudo
University of New Mexico
Robert Lawrence
Oregon State University
Don Layton
Cerritos College
Peter Leavens
University of Delaware
Barbara J. Leitner
University of Montevallo
John D. Longshore
Humboldt State University
Stephen J. Mackwell
Pennsylvania State University
Peter Martini
University of Guelph
G. David Mattison
California State University, Chico
George Maxey
University of North Texas
Lawrence D. Meinert
Washington State University, Pullman
Robert D. Merrill
California State University, Fresno

Kula C. Misra
University of Tennessee, Knoxville

Roger D. Morton
University of Alberta

J. Nadeau
Rider University

Peggy A. O'Day
Arizona State University

Terrence M. Quinn
University of South Florida

Simon M. Peacock
Arizona State University

E. Kirsten Peters
Washington State University, Pullman

Donald R. Prothero
Occidental College

C. Nicholas Raphael
Eastern Michigan University

J. H. Reynolds
West Carolina University

Robert W. Ridkey
University of Maryland

James Roche
Louisiana State University

William F. Ruddiman
University of Virginia

Charles K. Scharnberger
Millersville University

James Schmitt
Montana State University

Fred Schwab
Washington and Lee University

Steven C. Semken
Navajo Community College

D. W. Shakel
Pima Community College

Charles R. Singler
Youngstown State University

David B. Slavsky
Loyola University of Chicago

Douglas L. Smith
University of Florida

Richard Smosma
West Virginia University

Donald K. Sprowl
University of Kansas

Don Steeples
University of Kansas

Randolph P. Steinen
University of Connecticut

Bryan Tapp
University of Tulsa

John F. Taylor
Indiana University of Pennsylvania

Kenneth J. Terrell
Georgia State University

Thomas M. Tharp
Purdue University

Nicholas H. Tibbs
Southeast Missouri State University

Herbert Tischler
University of New Hampshire

Jan Tullis
Brown University

Kenneth J. Van Dellen
Macomb Community College

Lorraine W. Wolf
Auburn University

Peter W. Sloss (National Oceanic and Atmospheric Administration–National Environmental Satellite, Data, and Information Service–National Geophysical Data Center) created the digital images of Earth's crustal topography used throughout the book.

We have also had the benefit of informal advice on content and checking for accuracy from many colleagues, including especially C. W. Burnham, S. B. Jacobsen, Jane Selverstone, J. B. Thompson, Jr., Sean Solomon, Peter Molnar, Tom Jordan, George Wetherill, Steve Shirey, and Paul Silver.

Michael F. Follo (Colby College) contributed to the second edition in countless ways and supplied the boxed feature in Chapter 24. Dallas Rhodes (Whittier College) contributed the Internet Sources. Jill S. Schneiderman (Vassar College) contributed the Team Projects.

Others have worked with us more directly in writing and preparing manuscript for publication. At our side always were the editors at W. H. Freeman and Company, with Anne Vinnicombe and Nancy R. Fleming. The superb photographs that illuminate the text were obtained by Alexandra Truitt, who tirelessly combed many sources looking for the best possible choices.

The quality of the finished book would not have been possible without the final, and most skillful, efforts of Vertigo Design; Susan Wein, illustration coordinator; and Paul Rohloff, production coordinator. We are especially indebted to our illustrators—Ian Warpole, Tomo Narashima, and Network Graphics—for transforming our often very rough sketches into many outstanding drawings.

To the Student

Geology fascinates and excites us. We wrote *Understanding Earth* to help you discover for yourselves how interesting geology is in its own right and how important an understanding of geology has become for making decisions of public policy. What can we do to protect people and property from natural disasters such as volcanoes, earthquakes, and landslides? How can we use the resources of Earth—coal and oil, minerals, water, and air—in ways that minimize damage to the environment? In the end, understanding Earth helps us understand how to preserve life on Earth.

People tend to enjoy what they do well, and we want you to enjoy your geology course. Here are some tips for using our book in ways that will help you do well.

- **Chapter goals** Pay particular attention to the paragraph in each chapter that begins "This chapter will." This statement of goals will give you a firm idea of what the chapter covers and why.
- **Photographs and diagrams** Much of the information in geology is conveyed through illustrations. Our photographs are carefully chosen and diagrams carefully drawn to convey essential ideas. Whenever you see a reference to a figure number, examine the illustration thoroughly and read the caption. Reviewing the figures is a good way to refresh your memory of the chapter before class and before exams.
- **Boxed features** These typically are self-contained stories on topics of interest that reinforce the major themes of the book.
- **Summaries and Key Terms** The summary at the end of each chapter provides a concise description of the topics covered. Key terms and concepts are highlighted in boldface within the chapter, where they are first explained, and are listed with page references at the end of the chapter to help you review and test your understanding.
- **Exercises, Thought Questions, and Team Projects** These focus on important aspects of the chapter. Your instructor might select some of these for written homework or use them as a basis for test questions. Reviewing them quickly will help you test your memory and understanding of the chapter.
- **Suggested Readings** This list of books and journal articles will guide you to information on topics that interest you.
- **Internet Sources** Gain access to the latest research, visual materials, and more by trying the Web sites listed at the end of most chapters.
- **Study Guide** This learning aid features a wide variety of exercises to reinforce your knowledge of concepts and processes, self-tests, and suggestions for using the CD-ROM as a study tool.
- **CD-ROM** Your study and enjoyment of this book will be expanded to another dimension with this useful CD-ROM. Interactive exercises, practice quizzes, and a variety of electronic tools and resources have been designed to involve and engage you as you explore the key ideas of the textbook.
- ***Understanding Earth* Web Site** Updated regularly, this resource provides new information, new images, self-quizzes, and other helpful features. The address is: www.whfreeman.com/understandingearth

We hope you will find geology both intellectually satisfying and of practical value in preserving Earth for yourself and for future generations.

part 1

Understanding the Earth System

Our planet works as an interacting system of matter and energy that generates volcanoes, glaciers, mountains, lowlands, continents, and oceans. The energy that drives the system comes from the Earth's internal heat, which is responsible for plate tectonics, and solar radiation, which circulates the atmosphere and oceans and powers erosion. The matter of Earth—its rocks and minerals—and its structure are the relics of Earth system dynamics evolving over 4.6 billion years of geologic time. The three great clans of rocks and their geologic structures reflect geologic processes. Igneous rocks are linked to volcanism, sedimentary rocks to weathering and erosion, and metamorphic rocks to mountain building.

1

Aerial view of the Himalayas from Nepal. Mt. Everest, the world's highest peak, is on the right. (*Paul Steel/ The Stock Market*)

Building a Planet

Earth is a unique place, home to more than a million life forms, including ourselves. No other planet yet discovered has the same delicate balance of conditions necessary to sustain life. **Geology** is the science that studies Earth—how it was born, how it evolved, how it works, and how we can help preserve it.

This chapter gives a broad picture of how geologists think. It starts with the scientific method, the objective approach to the physical universe on which all scientific inquiry is based. Throughout this book, you will see the scientific method in action as you discover how geologists gather and interpret information about our planet. The chapter then describes the most generally accepted scientific explanations for how Earth formed and why it continues to change.

You may find as you read these pages that your idea of time will start to change. A geologist's view of time must accommodate spans so large that our minds sometimes have trouble comprehending them. Geologists estimate that Earth is 4.5 billion years old. About 3.5 billion years ago living cells developed on Earth, but our human origins date back only a few million years—a mere few hundredths of 1 percent of Earth's existence. The scales that measure individual lives in decades and mark off periods of written human history in hundreds or thousands of years are inadequate as we study Earth. Geologists must explain features that evolve over tens of thousands, hundreds of thousands, or millions of years.

THE SCIENTIFIC METHOD

The goal of all science is to explain with increasing precision how the universe works. The **scientific method,** on which all scientists rely, is a general research strategy based on the principle that every physical event has a physical explanation, even if it may be beyond our present ability to discover.

When scientists propose a **hypothesis**—a tentative explanation based on data collected through observations and experiments—they present it to the community of scientists for criticism and repeated testing against new data. A hypothesis that is confirmed by other scientists gains credibility, particularly if it predicts the outcome of new experiments.

A hypothesis that has survived repeated challenges and accumulated a substantial body of support is elevated to the status of a **theory.** Although its explanatory and predictive power has been demonstrated, a theory can never be considered finally proved. The essence of science is that no explanation, no matter how believable or appealing, is immune to question. If convincing new evidence indicates that a theory is wrong, scientists may modify or discard it. The longer a theory holds up to all scientific challenges, however, the more confidently it is held.

To encourage the atmosphere of challenge, scientists share their ideas and data by presenting them at professional meetings, publishing them in professional journals, and discussing them in informal conversations with colleagues. Scientists learn from one another's work as well as from the discoveries of the past. Most of the great concepts of science, whether they emerge as a flash of insight or in the course of painstaking analysis, result from untold numbers of such interactions. Albert Einstein said it this way: "In science . . . the work of the individual is so bound up with that of his scientific predecessors and contemporaries that it appears almost as an impersonal product of his generation."

Because such free intellectual exchange is subject to abuses, a code of ethics has evolved among scientists. Scientists must acknowledge the contributions of all others on whose work they have drawn. They must not falsify data. And they must accept responsibility for training the next generation of researchers and teachers.

THE PRINCIPLE OF UNIFORMITARIANISM

Much of what we have come to understand about the geologic past is based on observation of the workings of our planet today. We can observe today the growth of continents, the erosion of mountains, the eruptions of volcanoes. A fundamental principle of geology, advanced in the eighteenth century by the Scottish physician and geologist James Hutton, is that "the present is the key to the past." This **principle of uniformitarianism,** as it is now known, holds that the geologic processes we see in operation as they modify Earth's crust today have worked in much the same way over geologic time.

The rates of geological processes vary over a wide range. Continents can drift apart and mountains can be raised over millions of years. A large meteorite or comet can strike Earth and gouge out a vast crater, volcanoes can blow their tops, and earthquakes can rupture Earth's surface in seconds. These are events in an ongoing system that has continued to shape and reshape Earth since its birth 4.5 billion years ago (Figure 1.1).

Uniformitarianism, together with the laws of physics and chemistry, provides the basis for the theory and practice of geology. We will call upon uniformitarianism frequently as we attempt to decipher Earth's geologic history.

THE ORIGIN OF OUR SYSTEM OF PLANETS

The search for the origins of the universe and our own small part of it goes back to the earliest recorded mythologies. Today the most generally accepted scientific explanation is the "Big Bang" theory, which holds that our universe began some 10 billion to 15 billion years ago with a cosmic "explo-

FIGURE 1.1 Geologic phenomena can stretch over thousands of centuries or can occur with dazzling speed. (left) Mount Kerkeslin in the Canadian Rocky Mountains, Alberta, Canada. This mountain range was raised and deformed over a period of tens of millions of years. *(Carr Clifton.)* (right) Meteor Crater, Arizona. The explosive impact of a meteorite about 25,000 years ago excavated this crater in a few seconds. *(John Sanford/Photo Researchers.)*

sion." Before that moment, all matter and energy were compacted into a single, inconceivably dense point. Although we know little of what happened in the first fraction of a second when time began, astronomers have acquired a general understanding of the billions of years that followed. During that time, in a process that still continues, the universe has expanded and thinned out to form the galaxies and stars. Geologists focus on the last 4.5 billion years of this vast expanse, when our solar system—the star we call the Sun and the planets revolving around it—formed and evolved. Specifically, geologists look to the formation of the solar system in order to understand the formation of Earth.

The Nebular Hypothesis

In 1755 the German philosopher Immanuel Kant suggested that the origin of the solar system could be traced to a rotating cloud of gas and fine dust. Discoveries made in the past few decades have led astronomers back to this old idea, now called the **nebular hypothesis.** Equipped with modern telescopes, they have found that outer space beyond our solar system is not as empty as it once was thought to be. Astronomers have recorded many clouds of the type that Kant surmised, and they have named them *nebulae*. They also have identified the materials that form these clouds. The gases are mostly hydrogen and helium, the two elements that make up all but a small fraction of our Sun. The dust-sized particles are chemically similar to materials found on Earth.

How could our solar system take form from such a cloud? Part of the answer lies in the force of gravity, the attraction between pieces of matter because of their mass. This diffuse, slowly rotating cloud contracted under the force of gravity (Figure 1.2[a]). Contraction in turn accelerated the rotation of the particles (just as ice skaters spin more rapidly when they pull in their arms), and the faster rotation flattened the cloud into a disk (Figure 1.2[b]).

THE SUN FORMS Under the pull of gravity, matter began to drift toward the center, accumulating into a proto-Sun, the precursor of our present Sun

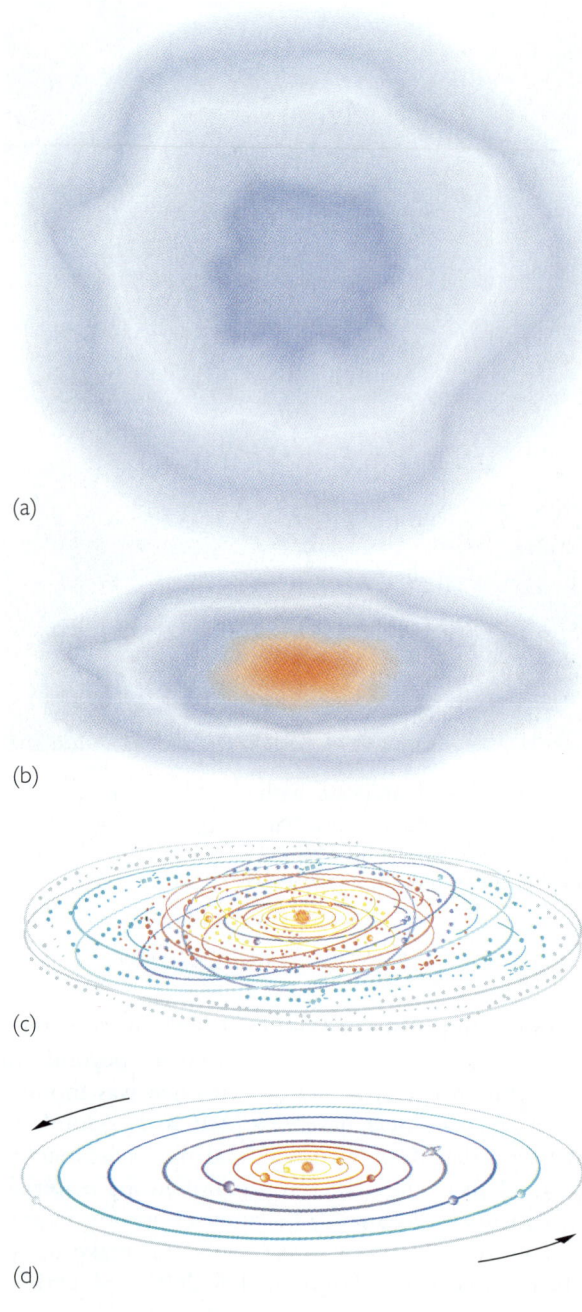

FIGURE 1.2 Evolution of the solar system. (a) A diffuse, roughly spherical, slowly rotating nebula begins to contract. (b) As a result of contraction and rotation, a flat, rapidly rotating disk forms with matter concentrated at the center that will become the proto-Sun. (c) The enveloping disk of gas and dust forms grains that collide and clump together into small chunks, or planetesimals. (d) The terrestrial planets build up by multiple collisions and accretion of planetesimals by gravitational attraction. Accretion of gas by the giant outer planets is not shown.

(Figure 1.2[c]). Compressed under its own weight, the material in the proto-Sun then became dense and hot. The internal temperature of the proto-Sun rose to millions of degrees, at which point nuclear fusion began. The Sun's nuclear fusion, which continues today, is the same nuclear reaction that occurs in a hydrogen bomb. In both cases hydrogen atoms, under intense pressure and at high temperature, combine (fuse) to form helium. Some mass is converted into energy in the process. This conversion is represented by Albert Einstein's famous equation $E = mc^2$, where E is the amount of energy released by conversion of mass (m) and c is the speed of light. Because c is a very large number (about 300,000 km per second) and c^2 is huge, a small amount of mass can yield an enormous amount of energy. Some of that energy is released as sunshine in the case of the Sun, and as a great explosion in the case of the H-bomb.

THE PLANETS FORM Although most of the matter in the original nebula was concentrated in the proto-Sun, a disk of gas and dust, called the solar nebula, remained to envelope it. The solar nebula grew hot as it flattened into a disk, hotter in the inner region, where more of the matter accumulated, than in the less dense outer regions. Once formed, the disk began to cool and many of the gases condensed. That is, they changed to their liquid or solid form, just as water vapor condenses into droplets on the outside of a cold glass and water solidifies into ice when it cools below the freezing point. Gravitational attraction caused the dust and condensing material to collide and accrete (clump together) as small chunks, or *planetesimals*. As the planetesimals collided and stuck together, larger Moon-size bodies formed (Figure 1.2[c]). In a final stage of cataclysmic impacts, a few of these larger bodies with their larger gravitational attraction swept up the others to form our nine planets in their present orbits (Figure 1.2[d]). Theoretical calculations indicate that all of this activity could have occurred in the remarkably short time of less than 100 million years. These rapid events occurred about 4.56 billion years ago, based on the age of meteorites that occasionally strike the Earth and that are believed to be remnants of that distant age.

As the planets formed, those in orbits close to the Sun and those in orbits farther from the Sun developed in markedly different ways.

- *The Inner Planets* The four inner planets, closest to the Sun, are Mercury, Venus, Earth, and Mars (Figure 1.3). They are also known as the terrestrial ("Earthlike") planets. In contrast to the outer planets, the four inner planets are small and rocky. They grew where conditions were so hot that volatile materials—those that become gases and boil away at relatively low temperatures—could not be retained in quantity. Radiation and matter streaming from the Sun blew away most of their hydrogen, helium, water, and other light gases and liquids, leaving behind dense metals such as iron and other heavy, rock-forming substances. By about 4.5 billion years ago, the inner planets had emerged as dense, rocky masses.

- *The Giant Outer Planets* According to this same scenario, most of the volatile materials swept from the region of the terrestrial planets were carried to the cold outer reaches of the solar system. Some accumulated on the giant *outer planets*—Jupiter, Saturn, Uranus, and Neptune—and their satellites. The rest were carried into outer space beyond. The giant planets were big enough and their gravitational attraction was strong enough to enable them to hold on to the lighter nebular constituents. Thus, although they have rocky cores, like the Sun, they are composed mostly of hydrogen and helium and the other light constituents of the original nebula. Tiny Pluto, orbiting farthest from the Sun, is a strange frozen mixture of gas, ice, and rock.

Testing the Nebular Hypothesis

This standard model of solar-system formation should be taken for only what it is—a tentative

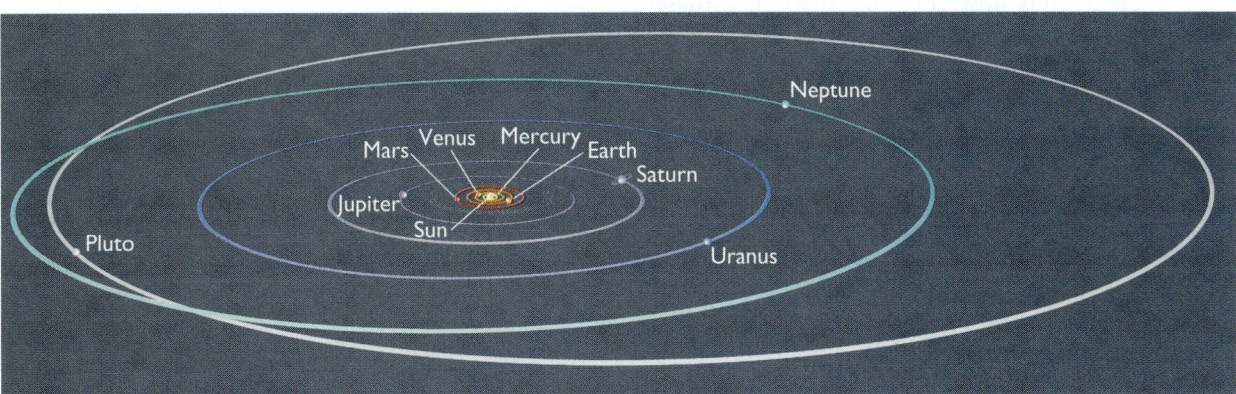

FIGURE 1.3 The solar system. The four inner planets—Mercury, Venus, Earth, and Mars—are closest to the Sun and are small and rocky. The four giant outer planets—Jupiter, Saturn, Uranus, and Neptune—and their satellites are mostly gaseous with rocky cores. The outermost planet, Pluto, is a snowball of methane, water, and rock. The upper panel shows the planetary orbits drawn roughly to scale; the distance from Pluto to the Sun averages about 5.9 billion km. The lower panel shows the relative sizes of the planets and the asteroid belt separating the inner and outer planets.

explanation that many scientists think best fits the known facts. Perhaps the model comes close to what actually happened. More important, however, is that this model gives us a way to think about the origin of the solar system. Scientists can examine its parts and modify it if necessary as more evidence comes in from new research. Such evidence has been gathered by American and Russian spacecraft carrying planetary probes. They have returned data on the nature and composition of the atmospheres and surfaces of Mercury, Venus, Mars, Jupiter, Saturn, Uranus, Neptune, and the Moon. A startling finding, which poses a problem for scientists who had hoped to learn how the planets evolved by comparing them, is that in our solar system—consisting of 9 planets and about 60 satellites—no two bodies are the same! Planetary scientists are searching for an answer.

Nebulae at various stages of development are being studied with powerful telescopes, and these observations are providing information about what goes on in the remote sections of the universe. A major discovery—made possible by the Hubble Space Telescope, which travels in orbit around Earth—was the first direct evidence of envelopes of gas, dust, and planets around several nearby stars. This finding strongly supports the idea that systems of planets in orbit around a sun also probably occur elsewhere in the universe. As the findings from these and other scientific investigations modify our working model, a clearer picture of the origin of our own solar system should emerge.

We have dwelt on the question of the origin of our solar system because the initial state of a planet helps determine its evolutionary course. Earth's current state is reasonably well known to us. The state of Earth at these two points in time, separated by some 4.5 billion years, must be fitted into any explanations we develop for the changes Earth has undergone throughout the course of its planetary evolution.

EARTH AS AN EVOLVING PLANET

How did Earth evolve from a rocky mass to a living planet with continents, oceans, and an atmosphere? The answer lies in **differentiation**—the transformation of a random mix of chunks of matter to a body in which the interior is divided into concentric layers that differ from one another both physically and chemically. Differentiation occurred early in Earth's history, when the planet got hot enough to melt.

The Earth Heats Up and Melts

To understand Earth's present layered structure, we must mentally return to the time when Earth was still subject to violent impacts by planetesimals and larger bodies. A moving body carries large amounts of kinetic energy, or energy of motion. (Think of how the energy of motion crushes a car in a collision.) A planetesimal colliding with Earth at a velocity of about 11 km per second would deliver as much energy as the same weight of TNT. When planetesimals and larger bodies crashed into the primitive Earth, most of this energy of motion was converted to heat, which is another form of energy. The impact energy of a Mars-sized body in collision with Earth would be equivalent to a trillion 1-megaton nuclear explosions, enough to eject into space a vast amount of debris and to deposit enough heat to melt most of what remained of Earth.

Many scientists now think such a cataclysm did indeed occur. Not only did it cause extensive melting, it also gave us the Moon. The giant impact propelled into space a shower of debris from both Earth and the impacting body, and the Moon aggregated from this debris (Figure 1.4). It would have reformed as a largely molten body. This huge impact also knocked the spin axis of the Earth from vertical

FIGURE 1.4 Artist's rendering of the collision of a Mars-size body with Earth about 4.5 billion years ago. The impact energy would have caused extensive melting of Earth and would have ejected debris that aggregated to form the Moon. (Painting by Alfred T. Kamajian, *Scientific American,* July 1994, cover.)

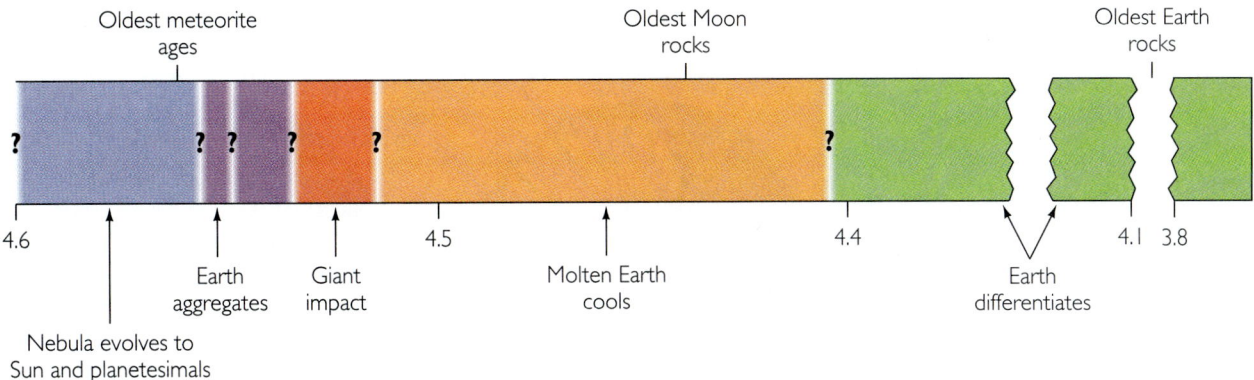

FIGURE 1.5 This timeline shows the origin of the Sun and Earth, the giant impact that melted much of the Earth and created the Moon, and the beginning of Earth's differentiation. The ages of the events listed above the bar are based on actual datings of meteorites and of lunar and Earth rocks. The question marks within the bar indicate a high degree of uncertainty about the timing of the events so marked. (Extensively modified from D. J. DePaolo, "Strange Bedfellows," *Nature,* Vol. 10, November 10, 1994, p. 131.)

with respect to the plane of the orbit to its present 23° inclination and sped up Earth's rotation. The oldest Moon rocks brought back by the *Apollo* astronauts are dated at 4.44 billion years ago, which should be close to the time of this violent event. If this hypothesis is correct, Earth's age can be bracketed between the age of meteorites, 4.56 billion years, and the age of the oldest Moon rocks.

In addition to the giant impact, another source of heat would have caused melting early in Earth's history. Several elements (uranium, for example) are radioactive. Although these elements occur in only tiny amounts, their radioactivity has had a profound effect on Earth's evolution. Atoms of radioactive elements spontaneously disintegrate by emitting subatomic particles. As these particles are absorbed by the surrounding matter, their energy of motion is transformed into heat. Radioactivity-generated heat, in addition to meteorite impacts, would have contributed to warming and melting in the young Earth. Radioactivity is a long-lived source of heat that continues to keep the interior warm.

Differentiation Begins

Early melting of Earth began the process of differentiation, perhaps the most significant event in Earth's history. It led to the formation of Earth's crust and eventually of the continents. It brought lighter materials to the outer layers of Earth and initiated the escape of even lighter gases from the interior, which eventually led to the formation of the atmosphere and oceans. Even today, gases continue to escape in the emissions that accompany volcanic eruptions. All of this activity began during the early melting of Earth. Figure 1.5 depicts as a timeline the major events in Earth's early history.

EARLY MELTING Although Earth probably began as an unsegregated mixture of the planetesimals and other remnants of the solar nebula, it did not retain this form long. Large-scale melting would have occurred as a result of the giant impact. Some workers in the field speculate that 30 to 65 percent of Earth melted, forming an outer layer hundreds of kilometers thick, which they call a "**magma** (molten rock) ocean." The interior, too, would have heated to a "soft" state in which its components could move around—heavy material sinking to the interior and lighter material floating toward the surface. The rising lighter matter would bring interior heat to the surface, where it could radiate into space. In this way Earth cooled and mostly solidified and was transformed into a differentiated or zoned planet with three main layers: a central core and an outer crust separated by a mantle (Figure 1.6).

EARTH'S CORE Iron, which accounted for about a third of the primitive planet's material, is denser than the other elements and sank to form most of a central **core.** Scientists have found that the core,

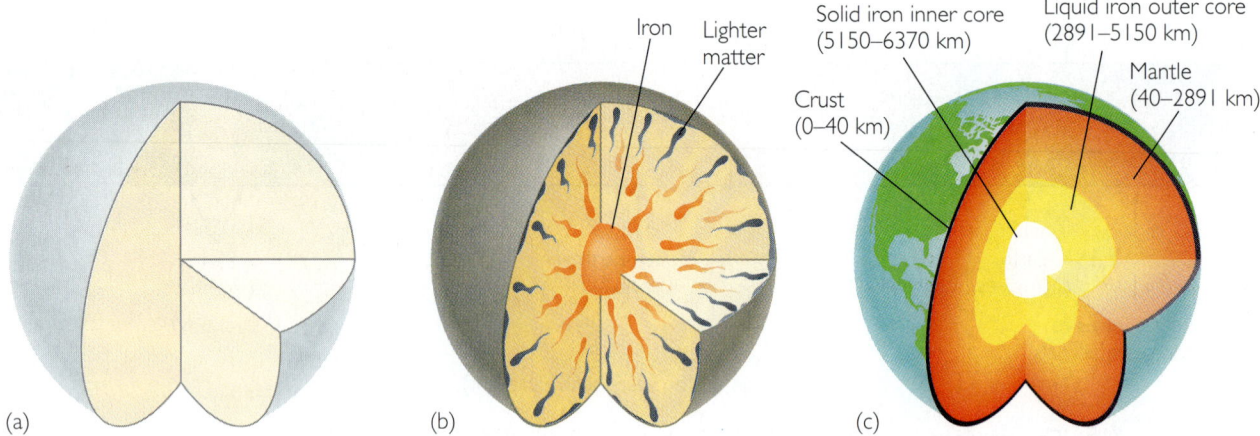

FIGURE 1.6 Early Earth (a) was probably a homogeneous mixture with no continents or oceans. In the process of differentiation, iron sank to the center and light material floated upward to form a crust (b). As a result, Earth is a zoned planet (c) with a dense iron core, a crust of light rock, and a residual mantle between them.

which exists at a depth beginning at 2900 km, is molten on the outside, but the inner core, from about 5200 to 6400 km, is solid. The reason is that the temperature at which any material melts increases with pressure. The iron core is solid nearest to Earth's center, where pressures are the highest.

EARTH'S CRUST Other molten materials were less dense than the parent substances from which they separated, so they floated toward the surface of the magma ocean. There they cooled to form Earth's solid **crust,** a thin outer layer ranging up to about 40 km in thickness. The crust contains relatively light materials with low melting temperatures. These are mostly compounds of the elements silicon, aluminum, iron, calcium, magnesium, sodium, and potassium, combined with oxygen. All of these materials, other than iron, are among the lightest of the solid elements. (Chapter 2 discusses chemical compounds and the elements from which they form.)

EARTH'S MANTLE Between the core and crust is the **mantle,** a region that is the bulk of the solid Earth. The mantle is the material left in the middle zone after most of the heavy matter sank and the light matter rose toward the surface. The mantle ranges from about 40 to 2900 km in depth. It consists of rock of intermediate density, mostly compounds of oxygen with magnesium, iron, and silicon.

Chemical analysis of rocks indicates that Earth contains more than 100 chemical elements, but 99 percent of its mass is made up of only eight (see Figure 1.7). Furthermore, about 90 percent of Earth consists of only four elements: iron, oxygen, silicon, and magnesium. When we compare the relative abundance of elements in the crust with their abun-

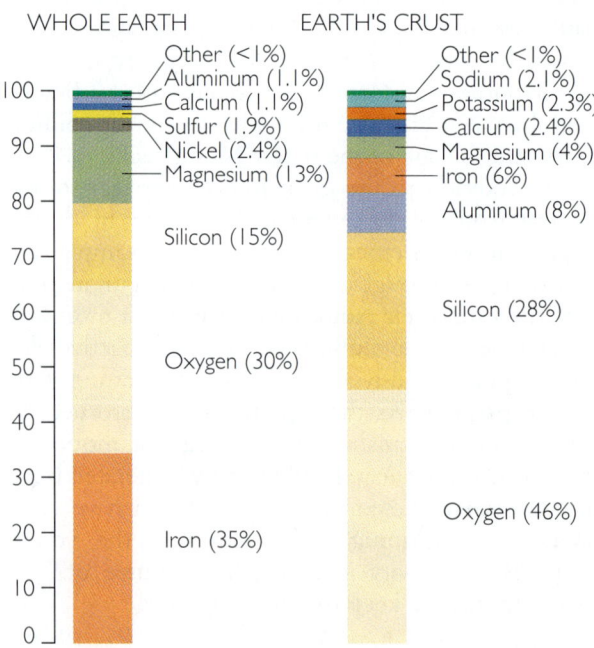

FIGURE 1.7 The relative abundance of elements in the whole Earth compared to that of elements in Earth's crust, given as percentages by weight. Differentiation has created a light crust, depleted of iron and rich in oxygen, silicon, aluminum, calcium, potassium, and sodium. Note that oxygen, silicon, and aluminum alone account for over 80 percent of the crust.

dance in the whole Earth, we find that iron accounts for a full 35 percent of Earth's mass, but because of differentiation there is little iron in the crust, where the light elements predominate. As you can see in Figure 1.7, the crustal rocks you stand on are almost 50 percent oxygen.

Earth's Continents, Oceans, and Atmosphere Form

Aside from the early, heat-producing impacts, from its very beginning Earth's history has been dominated by two constant heat engines, one internal, the other external. A heat engine—for example, the gasoline engine of an automobile—transforms heat energy released from fuel into mechanical motion or work. Earth's internal heat engine is powered by the heat generated by radioactivity. Earth's external heat engine is driven by solar energy—heat supplied to Earth's surface by the Sun. The internal heat melts rocks, forges volcanoes, and supplies the energy to build and move continents and to thrust mountains upward. The external heat is responsible for our climate and weather, and it drives the rain and wind that erode mountains and shape our landscape.

CONTINENTS Continental growth began soon after differentiation, and it has continued throughout geologic time. We have only the most general notion of what caused the formation of continents. We think magma floated up from the molten interior of Earth to the surface, where it cooled and solidified to form a crust of rock. This primeval crust melted and solidified repeatedly, allowing the lighter materials to separate gradually from the heavier ones and float to the top to form the primitive nucleus of the continents. Rainwater and other components of the atmosphere caused rocks to decompose and disintegrate. Water, wind, and ice then loosened and moved rocky debris to low-lying places, where it accumulated in thick layers, forming beaches, deltas, and the floors of adjacent seas. As this process was repeated through countless cycles, continents formed.

OCEANS AND ATMOSPHERE Most geologists believe that the origin of the oceans and atmosphere can be traced to Earth itself, that the oceans and atmosphere formed from water and gases that boiled off during heating and differentiation. A few other geologists propose an origin outside of Earth. Comets, they point out, are composed largely of water plus carbon dioxide, and other gases in frozen form. Countless comets may have bombarded Earth early in its history, carrying in water and gases that formed the early oceans and atmosphere.

Geologists who believe in an internal origin argue this way: Originally the water was locked up; that is, chemically bound as oxygen and hydrogen in certain minerals. Similarly, nitrogen and carbon were chemically bound in minerals. As Earth heated and its materials partially melted, water vapor and other gases were freed and carried to the surface by magmas and released through volcanic activity.

The gases released from volcanoes some 4 billion years ago probably consisted of the same substances that are expelled from present-day volcanoes: water vapor, hydrogen, carbon dioxide, nitrogen, and a few other gases (Figure 1.8). The earliest atmosphere thus was entirely different from the one we live in now, which consists primarily of nitrogen and oxygen. How did the atmosphere change? The production of significant amounts of free oxygen and its persistence in the atmosphere probably came about only after life had evolved at least to the complexity of photosynthetic algae, simple one-celled forms of life. Like other organisms that employ photosynthesis, algae use carbon dioxide and water as raw materials and the energy of sunlight to manufacture organic matter, and they release oxygen as a waste product. This oxygen began to accumulate in the atmosphere and gradually built up to its present concentration.

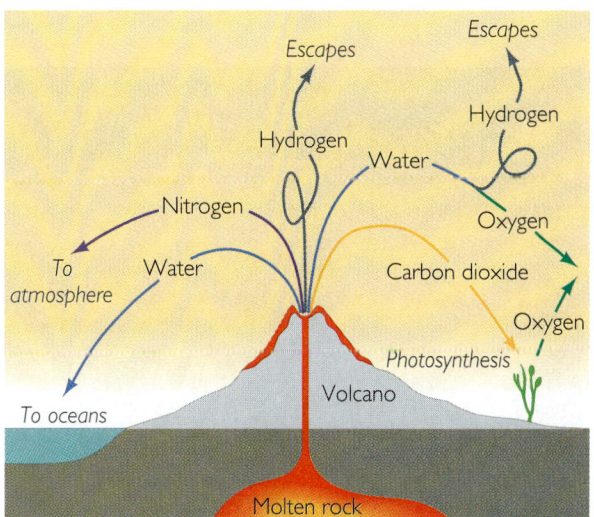

FIGURE 1.8 Volcanic activity has contributed enormous amounts of water, carbon dioxide, and other gases to the atmosphere and solid materials to the continents. Photosynthesis by plants removed carbon dioxide and added oxygen to the primitive atmosphere. Hydrogen, because it is light, easily escapes into space.

How the Other Planets and the Moon Evolved

By about 4 billion years ago Earth had become a fully differentiated planet. The core was still hot and mostly molten, but the mantle was fairly well solidified, and a primitive crust and continents had developed. Oceans and atmosphere had been produced, probably formed from substances released from Earth's interior, and the geologic processes we observe today were set in motion.

But what of the other planets? Did they go through the same early history? Information transmitted from our planetary spacecraft indicates that all the terrestrial planets have undergone differentiation, but their evolutionary paths have varied.

MERCURY Mercury, for example, has an iron core even larger than Earth's and a heavily cratered lightweight crust. Mercury had no blanket of air to protect it from being riddled by the impacts of large meteorites over the past 4.5 billion years. It is now geologically dead—that is, without the ongoing processes of mountain making, volcanic activity, and earthquakes that we observe on our own planet. It has a trace of an atmosphere but no wind or water to erode and smooth its ancient, cratered surface.

VENUS Venus is Earth's twin in mass and size, but it differs from Earth profoundly in the nature of its atmosphere and in its current geological behavior. Somehow—perhaps because its distance from the Sun is three-quarters that of Earth's, or because its early history of large impacts differed from that of Earth, or for some other reason—Venus evolved into a bone-dry planet. It is wrapped in a heavy, poisonous, incredibly hot (475°C) atmosphere, composed mostly of carbon dioxide and clouds of corrosive sulfuric acid droplets. Venus surpasses most descriptions of Hell: a human standing on its surface would be crushed by the pressure, boiled by the heat, and eaten away by the sulfuric acid. Or, as one planetary scientist exclaimed, "Earth is Earth, Venus is Venus; vive la différence!"

Because dense clouds shroud the surface of Venus, we knew relatively little about this terrestrial planet until recently. Now, radar images from the *Magellan* spacecraft, which went into orbit around Venus in 1990, suggest that the planet was geologically active in the past. The images show volcanoes, mountains, plateaus, plains, and other evidence of a dynamic surface (Figure 1.9). Yet its crust, unlike Earth's, today is relatively immobile. Something happened on Venus about 300 to 500 million years ago—perhaps a global outpouring of lava—that paved over the surface, obliterating many of the features of the first 85 percent of the planet's history. After this catastrophic event, the pace of geologic change seems to have slowed.

MARS Outermost of the terrestrial planets, Mars has a little over half the diameter of Earth. It has a

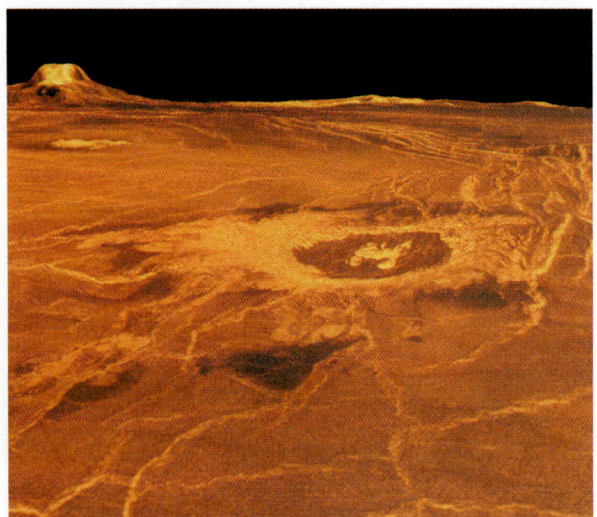 

FIGURE 1.9 Evidence of geological activity on Venus and Mars. (left) Venusian crater and volcano revealed by radar on the *Magellan* spacecraft, 1992. Crater in foreground, formed by meteorite impact, is 48 km in diameter. Volcano near horizon is 3 km high. *(NASA.)* (right) *Viking Orbiter* picture of the surface of Mars. Flat-floored valley may have been caused by water run-off early in the history of the planet. Large crater resulted from a more recent meteorite impact. *(Astrology Team, USGS.)*

crust, a core, and a composition similar to Earth's, and it has experienced many of the same geological processes that have shaped the Earth. Mars's surface shows that volcanic activity has occurred as recently as 10 million years ago and may still be going on. Craters, evidence of ancient meteorite impacts, are still preserved, in greater abundance than on Earth, where they have been mostly obliterated by subsequent geological activity. Although no liquid water is present on the surface today, networks of valleys and dry river channels indicate that liquid water was abundant on Mars before 3.5 billion years ago. This water may still be present as ice stored below the surface or sequestered in polar icecaps, or it may have evaporated and been lost to space.

Mars's thin atmosphere is composed almost entirely of carbon dioxide. It is unlikely that life exists on Mars today, but it is a serious question whether it could have existed in the past.

THE MOON The Moon is the best-known body in the solar system other than Earth because of its proximity and the programs of manned and unmanned exploration. As we discussed earlier, the favored theory of its formation proposes that it coalesced as a largely molten body after a giant impact ejected its matter from Earth. Analysis of lunar rocks brought back by the *Apollo* astronauts indicates that as this early magma ocean cooled and began to separate, lightweight, aluminum-rich compounds floated up to form a thick crust, and iron-rich materials sank to the interior. Thus the Moon underwent differentiation to form a small core and a thick crust, a process completed perhaps 4.35 billion years ago. In bulk, the Moon's materials are lighter than those of Earth because the heavier matter of the giant impactor and its primeval target remained embedded in the Earth.

The Moon has no atmosphere and, like Venus, is bone-dry, having lost its volatiles (substances that boil away at low temperatures) in the heat generated by the giant impact. The Moon cooled rapidly, and geological activity ceased some 3 billion years ago. The surface we see today is that of a very old, geologically dead body, with craters and mountains shaped almost entirely by the ancient impacts of large meteorites.

Evolution of the Outer Planets

The giant gaseous outer planets will remain a puzzle for a long time. They are so distinct chemically and so large that they must have followed an evolutionary course entirely different from that of the much smaller terrestrial planets. (It has even been proposed that Jupiter and its 15 moons are akin to a small solar system whose sun—Jupiter—never got quite massive enough for nuclear fusion to begin and then hot enough to shine.) Even less is understood about the evolution of Pluto.

Evidence of an Early Bombardment Period

The crater-pocked surface and the cratered surfaces of the Moon, Mars, Mercury, and other bodies are evidence of an important piece of the early history of the solar system—the **Heavy Bombardment period.** During this period, which may have lasted for 600 million years after the planets formed, the planets swept up and collided with the residual matter that had been left behind when they were assembled.

Every few million years a large chunk of matter still collides with Earth, sometimes with devastating effect. Most scientists believe that the impact of a large meteorite about 65 million years ago caused the extinction of many species, including the dinosaurs. In 1994, a once-in-a-millennium event occurred, when the comet Shoemaker-Levy crashed into Jupiter, exploding a plume of Jupiter's matter more than 1000 km above its atmosphere (Figure 1.10).

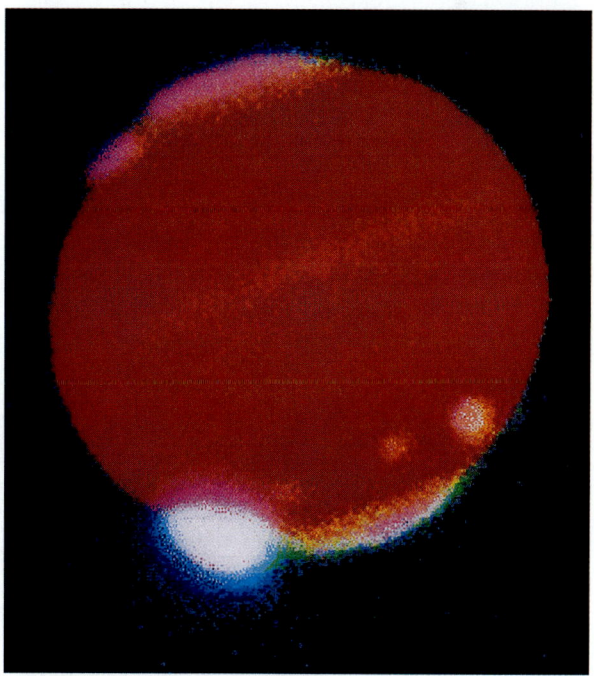

FIGURE 1.10 Infrared image showing the fireball created by the collision of the Comet Shoemaker-Levy with Jupiter on July 18, 1994. *(MSSSO, ANU/Photo Researchers.)*

Although such collisions have become extremely rare, some scientists have proposed that a number of telescopes should be assigned to search space and warn us months to years in advance of sizable bodies that might slam into Earth and devastate life over a large area.

PLATE TECTONICS: A UNIFYING THEORY FOR GEOLOGICAL SCIENCE

In the 1960s a great revolution in thinking shook the world of geologists. Physics had a comparable revolution at the beginning of the twentieth century, when the theory of relativity unified the physical laws that govern space, time, mass, and motion. Biology had a comparable revolution in the middle of this century when the discovery of DNA allowed biologists to explain how organisms transmit the information that controls their growth, development, and functioning from generation to generation. For almost 200 years geologists supported various theories of mountain building, volcanism, and other major phenomena of Earth, but no theory was general enough to explain well the whole range of geologic processes. We now have a single, all-encompassing concept that explains many of Earth's major geologic features. Furthermore, such topics as the classification and distribution of rocks and the positions and characteristics of volcanoes, earthquake belts, mountain systems, and ocean basins were formerly described more or less in isolation. Today we can treat these and other topics in the context of a unifying theory, **plate tectonics.** In the history of science, simple theories that explain many observations, as this one does, are the most enduring.

Plate tectonics is the idea that Earth's behavior is caused largely by the formation, movement, interactions, and destruction of the large rigid plates found at the surface of our planet. To understand what these plates are and how they were formed, we should return to our earlier discussion of Earth's crust and mantle and core.

We described the three main zones of Earth as chemically distinct layers that took their forms during differentiation. We could, however, also describe zones of Earth in terms of their physical properties. For example, you will see in Chapter 20 that Earth's zones can be characterized as strong and weak. We speak of "strong" here in the sense that a ceramic material is strong, and "weak" in the sense that modeling clay or wax is ductile, or plastic. The one is rigid and not easily deformed but can crack; the other is easily molded, like a tube of toothpaste. The **lithosphere** (from the Greek *lithos*, meaning "stone"), which includes the crust and the top part of the mantle, is depicted in Figure 1.11 as the strong, solid outermost shell, 50 to 100 km thick. The continents are raftlike inclusions embedded in the lithosphere. The lithosphere rides on the solid but weak **asthenosphere** (from the Greek *asthenes*, meaning "weak"). The lithosphere is strong because it is relatively cool, being so close to the surface. The asthenosphere, at greater depths, is a weak solid because it is hot, almost at the melting point.

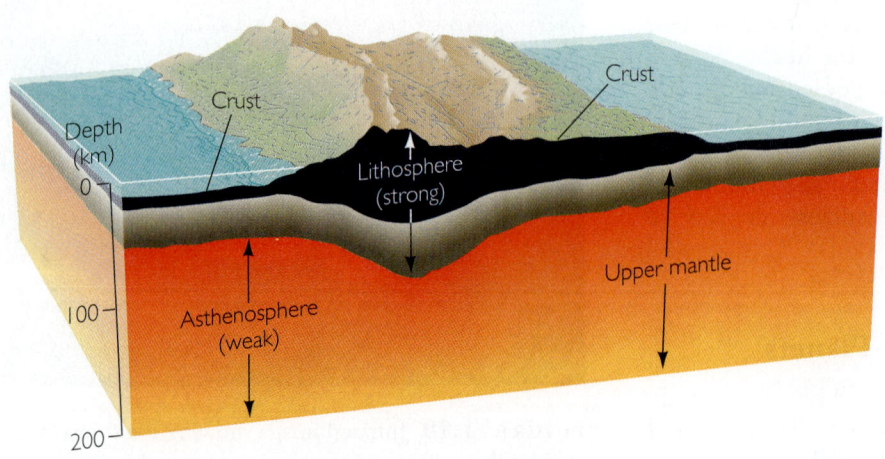

FIGURE **1.11** Earth's outermost shell is the strong, solid lithosphere, composed of the crust and top of the mantle. It rides on the weak, partially molten region of the mantle called the asthenosphere.

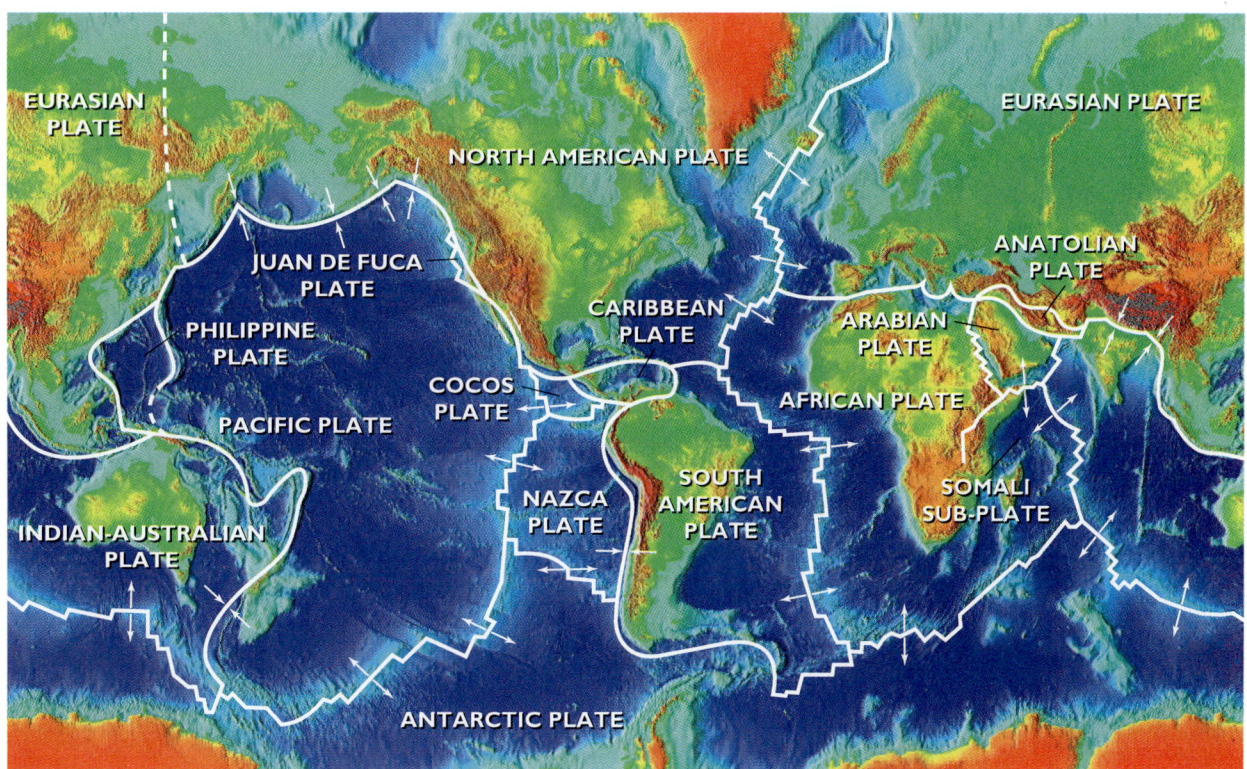

FIGURE 1.12 Earth's plates today. This flattened view of Earth's land and undersea topography shows plate boundaries, where plates separate (← →), collide (→ ←), or slide past each other (⇌). Note that plates and continents are not identical—the North American Plate, for example, is more extensive than the landmass that is the North American continent. *(Digital image by Peter W. Sloss, NOAA-NESDIS-NGDC.)*

The fact that the zones have different strengths determines that the lithosphere tends to behave as a rigid and brittle shell and the underlying asthenosphere "flows" as a ductile solid when it is subjected to forces.

Plates and Their Movements

According to the theory of plate tectonics, the lithosphere is not a continuous shell; it is broken into about a dozen large rigid plates that are in motion over the Earth's surface. Each plate moves as a distinct rigid unit, riding on the asthenosphere, which is also in motion. The major plates and the directions in which they move are shown in Figure 1.12. The North American Plate, for example, extends from the Pacific coast of the continent of North America to the middle of the Atlantic Ocean, where it meets the Eurasian and African plates.

Why should the plates move? Because Earth's interior is still hot. And even though the mantle beneath the lithosphere is mostly solid, it is hot and ductile or moldable. It can flow or "creep" if driven by forces. **Convection** supplies the forces. Convection is a mechanism of heat transfer that allows hot, less dense material to rise and dense, cool surface material to sink.

We tend to think of convection as a property of fluids and gases. We are all familiar with the circulating currents of boiling water in a pot, smoke rising from a chimney, heated air floating up to the ceiling, and cooled air sinking to the floor. Convection motion occurs in a flowing material, either a liquid or a moldable solid that is heated from below and cooled from above. The motion of solid flow is much slower than that of fluid flow. In either case, the heated matter rises under the forces of buoyancy because it has become less dense than the matter above it. It gives up heat and cools as it moves along the surface,

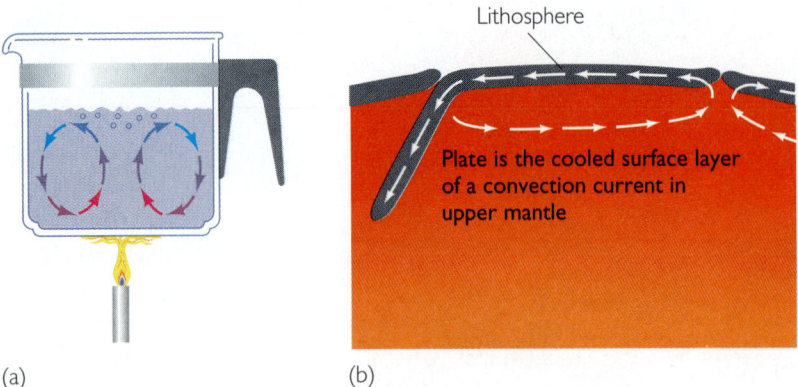

FIGURE 1.13 (a) In this familiar instance of convection, water rises as it heats and then falls again as it cools near the surface. (b) A simplified view of the way convection currents in the interior of Earth may be the reason why plates form and move. As hot matter rises under some plate boundaries, the plates form and diverge. At other boundaries where plates converge, a cooled plate sinks and is dragged under a neighboring plate. Note that continents are embedded in plates and move with them.

as depicted in Figure 1.13, becoming more dense. When it becomes "heavier" than the underlying material, it sinks under the pull of gravity. The circulation continues as long as heat remains to be transferred from the hot interior to the cool surface.

On Earth, the cooled surficial layer of the convective flow system becomes the rigid lithospheric plate. (Although the chemistry and convection system are entirely different, you might think of an ice sheet on a lake as a cooled surficial layer.) In this way the lithosphere forms from rising hot mantle where plates separate, cools as it moves away from this boundary, and sinks back into the asthenosphere, dragging the plate along. This process occurs at boundaries where plates converge.

The movement of plates at a rate of a few centimeters a year is the surface manifestation of Earth's convective system, still being driven by internal heat generated more than 4 billion years ago. This is a very simplified description of the mechanism of plate tectonics. Many details have been left out, such as the fate of the continents that are carried by the plates. We will be discussing these matters in due course. Although geologists generally agree on the "big picture," however, the details are still subjects of research and controversy.

Boundaries Between Plates

Many large-scale geologic features occur at the boundaries of plates, where the plates interact. These boundaries are of three types, all shown in Figure 1.14 and discussed in the following pages.

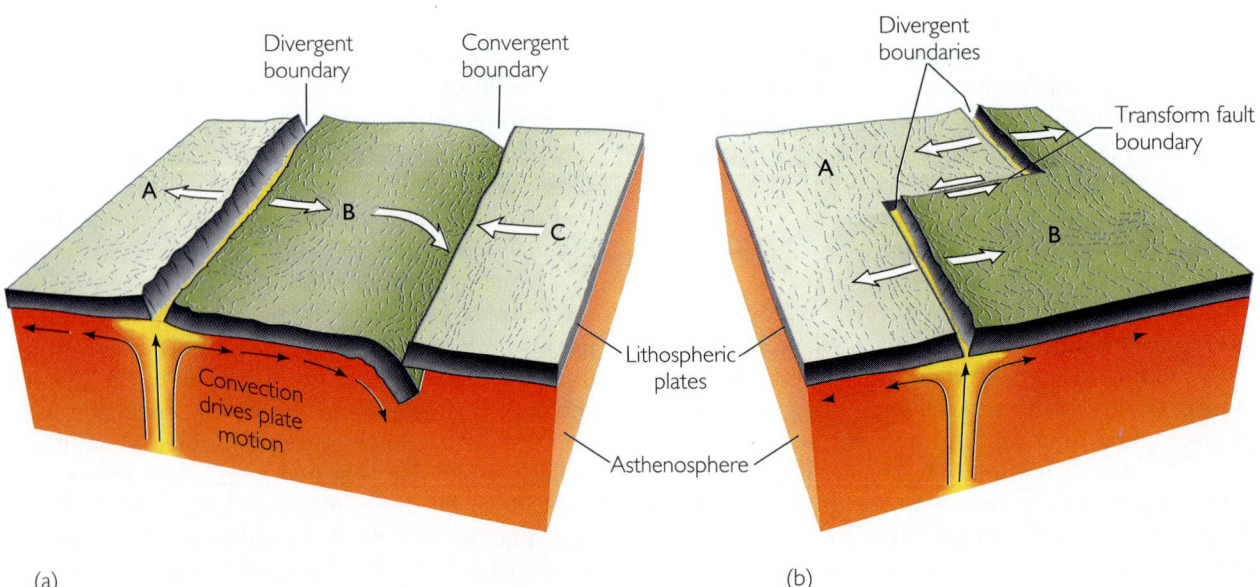

FIGURE 1.14 The three types of plate boundaries: (a) a divergent boundary, where plates A and B separate, and a convergent boundary, where plates B and C collide; (b) a transform fault boundary, where plates A and B slip past each other.

- **Divergent boundaries,** where plates separate and move in opposite directions, allowing new lithosphere to form from upwelling magma.
- **Convergent boundaries,** where plates collide and one sinks beneath the other, returning existing lithosphere to the interior.
- **Transform fault boundaries,** where plates slide past each other, approximately at right angles to their divergent boundaries.

DIVERGENT BOUNDARIES A divergent boundary is typified by a rift, or cracklike valley, at the crest of a **mid-ocean ridge,** a chain of volcanic mountains that winds along the bottom of the world's oceans. The Mid-Atlantic Ridge, for example, runs up the middle of the Atlantic Ocean and surfaces in several places, most extensively in Iceland (Figure 1.15). Other major divergent boundaries are the East Pacific Rise and the Indian Ocean ridges, which you can locate on the front endpapers.

FIGURE 1.15 The Mid-Atlantic Ridge, a plate divergence boundary, surfaces above sea level in Iceland. The cracklike valley indicates that plates are being pulled apart. *(Gudmundur E. Sigvaldason, Nordic Volcanological Institute.)*

Divergent boundaries are characterized by earthquakes and volcanoes as the plates move apart and magma rises in the space between them. The magma solidifies as rock in the crack between the plates, and the plates grow by the gradual accretion of this fresh rock. If plates continue to separate, new seafloor is created as ocean basins widen—a process called **seafloor spreading** (Figure 1.16).

CONVERGENT BOUNDARIES Because the plates cover the globe, if they separate in one place, they must converge somewhere else; and they do. Plates collide head-on along their convergent boundaries.

A profusion of geologic activities is associated with a plate collision. One plate sinks beneath the other, a process called **subduction** (Figure 1.16). Ocean lithosphere thus descends into the asthenosphere. This downbuckling produces a long, narrow deep-sea trench (about 100 km wide), where the ocean floor reaches its greatest depths (about 10 km below sea level). The edge of the overriding plate (in Figure 1.16, a plate with a continent on its edge) is crumpled and uplifted to form a mountain chain roughly parallel to the trench. The enormous forces of collision and subduction produce great earthquakes. Materials may be scraped off the descending slab and incorporated into the adjacent mountains. Imagine yourself as a geologist attempting to figure out the meaning of such tangled evidence. Furthermore, during subduction, parts of the descending plate may begin to melt. Magma formed where plates sink into the mantle floats upward, and can reach the surface and erupt from volcanoes.

Recall that divergent zones are sources of new lithosphere. **Subduction zones** at boundaries of convergence are sinks in which lithosphere is consumed by being returned to the mantle.

The west coast of South America, where the South American Plate collides with the oceanic Nazca Plate, is a subduction zone at a convergent boundary. (This and other convergent boundaries can be located on the front endpapers.) The Andes Mountains rise on the continental side of this boundary, and the Chilean deep-sea trench lies just off the coast. In this locale, volcanoes are active and deadly. One of them, Nevado del Ruiz in Colombia, was responsible for the deaths of 25,000 people when it erupted in 1985. Some of the world's greatest earthquakes have also been recorded along this boundary.

Another subduction zone is the boundary between the small Juan de Fuca Plate and the North American Plate, just off the coasts of British Columbia, Washington, and Oregon This convergent boundary gives rise to the volcanoes of the Cascade Range, including the dangerously active Mount St. Helens. As

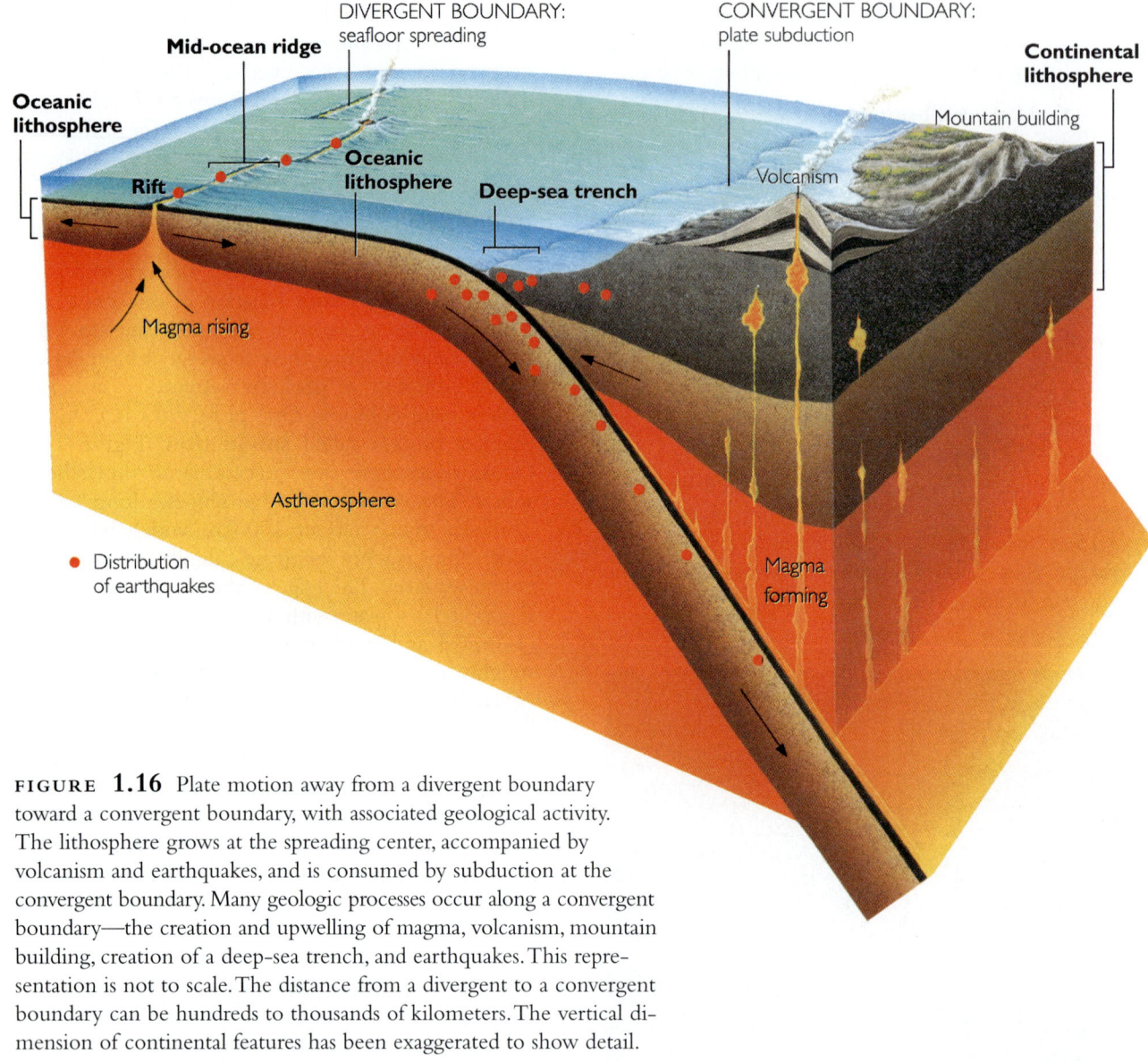

FIGURE 1.16 Plate motion away from a divergent boundary toward a convergent boundary, with associated geological activity. The lithosphere grows at the spreading center, accompanied by volcanism and earthquakes, and is consumed by subduction at the convergent boundary. Many geologic processes occur along a convergent boundary—the creation and upwelling of magma, volcanism, mountain building, creation of a deep-sea trench, and earthquakes. This representation is not to scale. The distance from a divergent to a convergent boundary can be hundreds to thousands of kilometers. The vertical dimension of continental features has been exaggerated to show detail.

our understanding of this subduction zone grows, scientists are increasingly worried that it will be the site of a future great earthquake affecting Oregon, Washington, and British Columbia.

TRANSFORM FAULT BOUNDARIES Some plates do not collide; they slip past each other horizontally along a transform fault. The famed San Andreas fault of California is such a boundary. There the Pacific Plate slides past the North American Plate in a northwesterly direction (Figure 1.17). Because the plates have been sliding past each other for millions of years, rocks facing each other on the two sides of the fault are of different types and ages. The sliding is not smooth, but more of a stick-slip process, in which sudden slips produce damaging earthquakes. One such earthquake destroyed San Francisco in 1906. There is much concern that a sudden slip along the San Andreas fault or related faults near Los Angeles or San Francisco may be extremely destructive within the next 25 years or so.

CHAPTER 1 • BUILDING A PLANET   19

Plate Tectonics and Planetary History

If continents are embedded in plates, it follows that the geography of the world must have been different in the remote past. The movement of continents over geologic time, termed **continental drift**, is an important part of modern geologic analysis. Some 200 million years ago all of the continents were assembled together into the supercontinent of Pangaea, as shown in Figure 1.18. (We discuss the geologic evidence for continental drift and Pangaea in Chapter 20.)

The plate motions we have just described represent the general pattern of the work output of Earth's internal heat engine as we can see it today. Because of Earth's internal heat, plates have emerged, moved, and been destroyed since the large-scale differentiation of Earth ended some 4 billion years ago. Geologists ponder why plate tectonics is not active today on the other terrestrial planets, but may have been in the past. In later chapters we will consider in more detail how a living planet generates and gets rid of its heat so that we can understand how that planet works.

The Theory of Plate Tectonics and the Scientific Method

Earlier we discussed the scientific method and the ways in which it guides the work of geologists. In the context of the scientific method, plate tectonics is not a dogma but a confirmed theory whose strength lies in its simplicity and generality. Theories can be overturned, but the theory of plate tectonics—like the theories of the age of Earth, the evolution of life,

FIGURE 1.17 The view southeast along the San Andreas fault in the Carrizo Plain of central California. The San Andreas is a transform fault, forming a portion of the sliding boundary between the Pacific plate on the right and the North American plate on the left. *(Michael Collier.)*

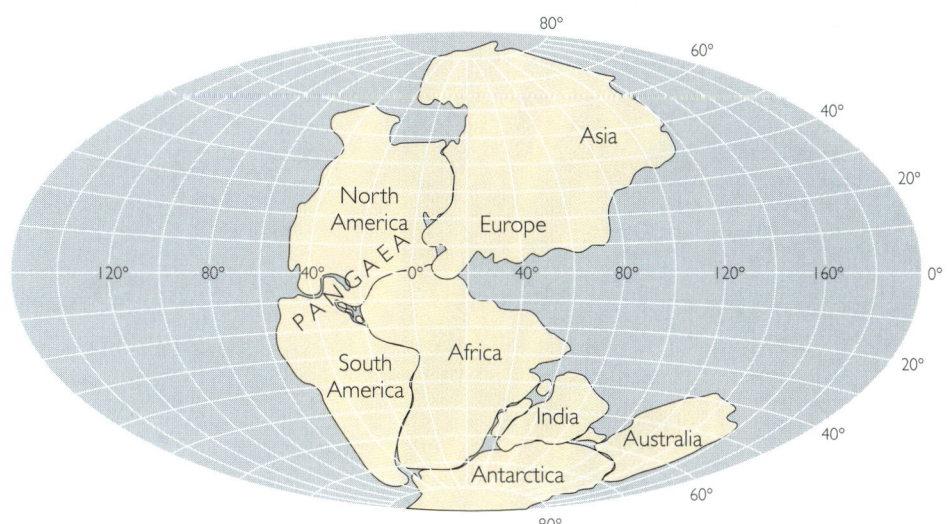

FIGURE 1.18 The supercontinent of Pangaea ("all lands"). Our current understanding of plate motions on the Earth leads us to believe that the geography of the world may have been different 200 million years ago. At that time, all of the continents were probably joined together as a single supercontinent, as shown here.

and genetics—explains so much so well and has survived so many efforts to prove it false that most geologists treat it as fact (see Feature 1.1).

Nevertheless, competing hypotheses have been advanced to explain the forces that cause plates to move, and others will probably be advanced in the future. Our story of the origin and early evolution of Earth is a hypothesis, and we can expect it to change many times because of the difficulty of recovering the information contained in the oldest rocks—information largely destroyed during the violent process of differentiation.

Because available information is incomplete and scientists' ability to observe nature is limited, some questions can be answered no better by scientists than by philosophers or poets. What, for instance, existed before the Big Bang? Or, as Walt Whitman asked:

Great is the Earth, and the way it became what it is,
Do you imagine it is stopped at this?

In this chapter we have made many statements without providing supporting observations or an underlying rationale. We have done so in order to preview the big picture, reserving for later chapters all of the substantiation called for by the scientific method. We began with the origin of the universe, but most of our discussion has dealt with the events that led to the formation and evolution of our planet to the point where continents, oceans, and an atmosphere developed. Now that we have outlined plate tectonics, the paradigm that has shaped geologic thought in the second half of the twentieth century, we can consider continents and ocean basins and interpret information contained in the various kinds of rocks within that conceptual framework. Although plate tectonics does not explain everything, it appears to be the best foundation on which to structure Earth's story.

GEOLOGISTS AT WORK

Although all geologists approach their work as scientists, they vary in the aspects of geology they embrace most avidly. Some are interested in geology as a pure science, a fascinating intellectual pursuit in its own right. Other geologists are more interested in using their knowledge in practical ways (Figure 1.19). They may join in the search for essential resources—the coal, oil, natural gas, metals, chemicals, and other materials on which modern civilization depends. They may warn of impending volcanic eruptions or help in the design of river flood control systems.

Scholarly Geology

Like most scientists, scholarly geologists seem to have an unbounded curiosity about the world around them, and they tend to feel uneasy when important natural phenomena remain unexplained. With the impulse of explorers, geologists will be hammering at rock outcrops, drawing up geologic maps, probing the seafloor, and scrutinizing Moon rocks as long as

FIGURE 1.19 Geologist collecting samples from a lava flow on Kilauea volcano, Hawaii.
(David R. Frazier/Photo Researchers.)

the workings of our planet remain incompletely explained. They may serve as teachers and researchers in universities, or work for government agencies or for private firms that engage in construction or draw materials from the Earth.

Mitigating Nature's Threats

Some geologists are primarily interested in advancing our understanding of processes that can result in natural disasters. As human populations multiply and become increasingly concentrated, our vulnerability to natural disasters also increases. With greater urgency than ever before we must seek safeguards against nature's threats: earthquakes and the destructive sea waves that often accompany them, volcanic eruptions, landslides, floods, and droughts.

Geologists have had some success in this ongoing search. We have learned, for example, that certain earthquakes below the seafloor can trigger giant waves with the potential to destroy communities on distant coasts. A warning system now provides people along the Pacific coast with a few hours' notice of the arrival of such waves. Systems are now in place in other countries to monitor volcanoes and predict eruptions so that authorities can evacuate endangered populations. Japan, China, Russia, and the United States are testing methods to predict impending earthquakes. And in many countries around the globe, geologists play a role in identifying areas threatened by landslides and floods so that they can be rezoned to prohibit construction and thus preclude later devastation (Figure 1.20).

PROTECTING THE ENVIRONMENT Of all nature's threats to our planet, humans may be the greatest. Our species has gained the power to foul the land, seas, and sky. Urbanization, mining, agricultural operations, and warfare now rival nature in their ability to modify Earth's surface.

Groundwaters, which are the source of much of our drinking water, are beginning to show worrisome levels of toxic chemicals. These substances originate from agricultural and industrial wastes that seep through the soil and rock to the water stored below.

Acid rain, in part a product of coal combustion and automobile exhaust, threatens not only our lakes but also our forests.

Earth's atmosphere has functioned like a greenhouse, raising Earth's temperatures to levels that make life possible. The combustion of coal and oil, along with other industrial activities and the destruction of forests, releases gases that may be intensifying the greenhouse effect. It now seems quite possible that our climate will grow warmer in 50 to 100 years as a result. Unchecked, this trend could in time convert some fertile agricultural regions to semiarid lands, and melting glaciers could raise the level of the seas until it floods some low-lying coastal cities. By studying the causes and effects of past natural climatic changes, geologists are providing important knowledge about the possible impact of human activities on the environment.

SEARCHING FOR UNDISCOVERED RESOURCES From early Stone Age times, humans have been

FIGURE 1.20 Some natural disasters can be avoided. Many buildings were destroyed when heavy rains caused this landslide in Hong Kong. It is too late to relocate this area of downtown Hong Kong, but in other areas geologists can help determine whether environmental risks to a proposed building site dictate special construction or no construction at all.

1.1 INTERPRETING THE EARTH

Continental Drift: A Case History of the Scientific Method at Work

The idea that continents can move over Earth's surface through geologic time has a 140-year history of hypotheses successively advanced, criticized, rejected, and modified and improved with new data.

Francis Bacon first noticed in 1660 the jigsaw-puzzle fit of Africa and South America. Not until the mid-nineteenth century, however, was the conformity of the two coastlines proposed as evidence of the breakup and separation of the two continents. Few supported this farfetched idea. There was evidence that continents had moved up and down through geologic time but none, as yet, that they could move across Earth's surface. Uniformitarianism's tenet that the present is the key to the past had become firmly implanted in geologic thought. The conventional view was that the continents were and therefore always would be anchored in their present spots.

Although several scientists in the following years offered additional observations in support of continental drift, it was the German meteorologist Alfred Wegener who in 1912 ventured outside his own specialty and proposed a hypothesis that could not be ignored.

From paleontology, Wegener found evidence that many fossil and living organisms are remarkably similar between North America and Europe, South America and Africa, Australia and India, South Africa and southernmost South America. From geology, he found that different types of rock formations matched up if the continents were reassembled. Wegener explained his data by postulating a supercontinent, "Pangaea," that broke apart about 200 million years ago (see Figure 1.18). The fragments, which became the continents we know today, moved apart, and oceans filled the growing voids between them. Wegener wrote: "It is just as if we were to refit the torn pieces of a newspaper by matching their edges and then check whether the lines of print run smoothly across. If they do, there is nothing left but to conclude that the pieces were in fact joined this way" (Hallam 1973).

Wegener's hypothesis found few supporters. Some scientists pointed out that the coastlines did not fit together precisely. They also questioned whether the rock formations on both sides of the Atlantic matched as Wegener claimed, and whether this necessarily implied that the continents once were joined.

Paleontologists offered an alternative hypothesis for the similarities in fossil plants and animals. A few land bridges across the ocean could have provided a path for the migration of plants and animals. If the land bridges subsided and disappeared beneath the sea 200 million years ago, there was no need to postulate the breakup of a supercontinent.

Others argued against Wegener's hypothesis because they thought that the rocks of Earth's outer layers were too stiff for continents to plow through them. Moreover, Wegener had not provided a believable physical

---

searching Earth's crust for sources of energy and materials to support their current level of culture and society. That search has depleted some of the most accessible supplies of Earth's resources. Gone are the days when prospectors could easily find deposits of oil, iron, copper, tin, uranium, and other materials important to the world's economies. But if we are to see life improve for people in less developed countries, let alone maintain our own present standard of living, new mineral deposits must be found. The challenge for economic geologists is to reexplore the world, using new tools and techniques to search out the remaining undiscovered deposits of coal, oil, natural gas, and the ores from which our metals and industrial chemicals are derived.

The exploitation of Earth's mineral wealth has raised new and pressing concerns for geologists, principally protection of the environment and conservation of resources. How can we most efficiently exploit nature's wealth without waste and without devastating the landscape? Somehow we must find answers to this question.

We hope that some students taking this geology course will choose careers in Earth sciences or environmental sciences and that all readers will develop a lifelong interest in understanding Earth.

mechanism to explain why continents should be forced to move at all.

Wegener himself admitted that his hypothesis was based on circumstantial evidence and incomplete data. His opppnents accused him of ignoring contradictory data. However, his ideas found some support from a few important geologists, who corrected some of Wegener's mistakes, found additional evidence, and began to suggest believable mechanisms to move continents. Still, it was not enough to elevate Wegener's hypothesis to a respectable theory. An important reason was that almost all their data came from the continents, with little from the unexplored seafloor which covers 80 percent of Earth's surface.

World War II saw the development of new types of scientific instruments for military purposes. When the wartime geologists returned to their civilian life they began to use many of these powerful tools in their peacetime oceanographic research. They were able to map the topography of the seafloor rapidly and study its rocks with remote sensors. They discovered the mid-ocean ridges and showed that the seafloor is progressively older away from the ridges, as should occur with seafloor spreading. They found no evidence of submerged land bridges across the Atlantic—a death knell for that hypothesis. With this and other evidence they could provide overwhelming support for seafloor spreading, continental drift, and plate tectonics.

Wegener's hypothesis now has a basis in firm and abundant data, and supporting evidence continues to accumulate. Global Positioning Satellites, which can locate any position on Earth to a centimeter or so, can show on a yearly basis that the Atlantic Ocean is opening at a rate as fast as your fingernails grow.

Wegener died in 1930 on an expedition on the Greenland icecap, never knowing that he would be regarded as the grandfather of the plate tectonics revolution. Geologists are still researching the mechanism that causes plates to move and continents to drift with them.

Globe generated from digital data bases of land and seafloor elevations clearly shows the mid-Atlantic Ridge, where seafloor spreading is moving the South American and African plates apart.
(Peter W. Sloss, NOAA-NESDIS-NGDC.)

Summary

What is geology? Geology is the science that deals with Earth—its history, its composition and internal structure, and its surface features.

How do geologists study Earth? Geologists, like other scientists, use the scientific method. They share the data they develop and check one another's work. A hypothesis is a tentative explanation of a body of data. If it is confirmed repeatedly by other scientists' experiments, it may be elevated to a theory. Many theories are abandoned when subsequent experimental work shows them to be false. Confidence grows in those theories that withstand repeated tests and are able to predict the results of new experiments. Geologists have confidence in the principle of uniformitarianism. They believe that the processes that have shaped Earth have not changed over geologic time, and that therefore the key to understanding the past lies in observing how those processes work today.

How did our solar system originate? The Sun and its family of planets probably formed when a primeval cloud of gas and dust condensed about 4.5 billion years ago. The planets vary in chemical composition in accordance with their distance from the Sun and with their size.

How did Earth form and evolve over time? Earth probably grew by accretion of colliding chunks of matter. Soon after it was formed, it was struck by a giant meteorite. The impact had a profound effect on our planet. Matter ejected into space from both Earth and the impactor reassembled to form the Moon. The impact also melted much of the Earth. Radioactivity also contributed to early heating and melting. Heavy matter, rich in iron, sank toward Earth's center, and lighter matter floated up to form the outer layers that became the crust and continents. Outgassing gave rise to the oceans and a primitive atmosphere. In this way Earth was transformed to a differentiated planet with chemically distinct zones: an iron core; a mantle that is mostly magnesium, iron, silicon, and oxygen; and a crust rich in the light elements oxygen, silicon, aluminum, calcium, potassium, and sodium and in radioactive elements.

What are the basic elements of plate tectonics? The lithosphere, Earth's outermost shell, is broken into about a dozen large, rigid plates. These plates and their origin and movement are a manifestation of solid-flow convection currents in the mantle. The plates jostle each other as they move in their individual courses. The boundaries between divergent and convergent plates are zones of intense activity: mountain building, volcanism, seafloor creation and destruction, and earthquakes. Major earthquakes occur on transform fault boundaries.

How is geology both a basic and an applied science? Geology is a basic science in that it creates new understanding and knowledge of the planet. It is an applied science in its involvement with discovering Earth's mineral wealth and mitigating hazards such as earthquakes, volcanoes, floods, and environmental damage.

KEY TERMS AND CONCEPTS

geology (p. 3)
scientific method (p. 4)
hypothesis (p. 4)
theory (p. 4)
principle of uniformitarianism (p. 4)
nebular hypothesis (p. 5)
differentiation (p. 8)
magma (p. 9)

core (p. 9)
crust (p. 10)
mantle (p. 10)
Heavy Bombardment period (p. 13)
plate tectonics (p. 14)
lithosphere (p. 14)
asthenosphere (p. 14)
convection (p. 15)

divergent boundaries (p. 17)
convergent boundaries (p. 17)
transform fault boundaries (p. 17)
mid-ocean ridge (p. 17)
seafloor spreading (p. 17)
subduction (p. 17)
subduction zones (p. 17)
continental drift (p. 19)

EXERCISES

1. What is the difference between an experiment, a hypothesis, a theory, and a fact?

2. What factors made Earth a particularly congenial place for life to develop?

3. How and why do the inner planets differ from the giant outer planets?

4. What caused Earth to differentiate, and what was the result?

5. How does the chemical composition of Earth's crust differ from that of its deeper interior? From that of its core?

6. Describe the central idea of the theory of plate tectonics.

7. Describe and explain the large-scale geologic activities associated with the three types of plate boundaries.

THOUGHT QUESTIONS

1. How does the discovery of solid matter around other stars contribute to the debate about the possibility of life elsewhere in the cosmos? What are the implications of the existence of life on the planets of other stars?

2. If you were an astronaut exploring another planet, how would you decide whether the planet was differentiated and whether it was still geologically active?

3. What are the advantages and disadvantages of living on a differentiated planet? On a geologically active planet?

4. If a giant impact occurred after life had formed on Earth, what would have been the consequences?

5. Speculate on what life would be like today if the ancient continent of Pangaea had remained a single landmass instead of breaking up into Eurasia, Africa, and the Americas.

6. According to biblical interpretation, Earth is some 5000 years old. Is that figure a hypothesis, a theory, a fact, or an article of faith? What is the difference between a statement based on faith and a scientific theory?

SUGGESTED READINGS

Ahrens, Thomas J. 1994. The origin of the Earth. *Physics Today* (August): 38–35.

Allegre, Claude. 1992. *From Stone to Star*. Cambridge, Mass.: Harvard University Press.

Brandt, J. C., and S. P. Maran. 1979. *New Horizons in Astronomy*, 2nd ed. San Francisco: W. H. Freeman.

Hallam, A. 1973. *A Revolution in the Earth Sciences: From Continental Drift to Plate Tectonics*. Oxford: Clarendon Press.

Hurley, P. M. 1968. The confirmation of continental drift. In *Continents Adrift, Readings from Scientific American*, 56-67. San Francisco: W. H. Freeman.

Exploring Space. 1990. Special issue of *Scientific American*.

Managing Planet Earth. 1989. Special issue of *Scientific American* (September).

May, Robert H. 1992. How many species inhabit Earth? *Scientific American* (April): 42–48.

National Academy of Sciences. 1992. *Science and Creationism*. Washington, D.C.: National Academy Press.

National Research Council. 1990. *The Search for Life's Origins*. Washington, D.C.: National Academy Press.

National Research Council. 1993. *Solid-Earth Sciences and Society*. Washington, D.C.: National Academy Press.

Press, Frank, and Raymond Siever. 1986. The planets: A summary of current knowledge. Chapter 22 in *Earth*, 4th ed. New York: W. H. Freeman.

Stanley, Steven M. 1993. *Exploring Earth and Life Through Time*. New York: W. H. Freeman.

Takeuchi, H., S. Uyeda, and H. Kanamori. 1970. *Debate About the Earth*. San Francisco: W. H. Freeman, Cooper & Co.

Taylor, G. Jeffrey. 1994. The scientific legacy of *Apollo*. *Scientific American* (July): 40–47.

Understanding Earth 2.0 CD-ROM. 1997. New York: W. H. Freeman.

Walter, William J. 1992. *Space Age*. New York: Random House.

Weiner, Jonathan. 1986. *Planet Earth*. New York: Bantam Books.

Westbroek, Peter. 1991. *Life as a Geologic Force*. New York: W. W. Norton.

Wetherill, George W. 1990. Formation of the Earth. *Ann. Rev. Earth Planet. Sci.* 18: 205–256.

INTERNET SOURCES

Ask-A-Geologist
❶ http://walrus.wr.usgs.gov/docs/ask-a-ge.html
Questions addressed to this site will be answered by a geologist with the U.S. Geological Survey. The site also includes a list of Frequently Asked Questions (FAQ).

The Geologist's Lifetime Field List
❶ http://www.uc.edu/~ACOMBTY/geologylist.html
This site provides a list of the features and events that most geologists would like to experience firsthand during their careers. Links are provided to more information about the items on the list.

Earth Science Site of the Week
❶ http://agcwww.bio.ns.ca/misc/geores/sotw/sotw.html
The Atlantic Division of the Canadian Geological Survey features an outstanding site dealing with a different aspect of earth science each week. A list of previous selections is available with links to the sites.

The Nine Planets
❶ http://www.seds.org/billa/tnp/
More than 60 "pages" of text and images dealing with the solar system are available here. You will find multiple images of the features comprising the solar system, video of a few features, a glossary, and a list of "Open Issues" about each member of the solar system. Take the "Quick Tour" for an overview of the site.

Views of the Solar System
❶ http://bang.lanl.gov/solarsys/homepage.htm
This site includes an image gallery from NASA and other sources. With links to pages for each component of the solar system, this site provides basic data, images, video, and more. The text is available in English and Spanish.

Welcome to the Planets
❶ http://pds.jpl.nasa.gov/planets/
This NASA site provides data and images for each of the planets, a glossary, and information on the spacecraft that produced the images.

Web Site for "Understanding Earth"
❶ www.whfreeman.com/understandingearth/
The publisher of *Understanding Earth* updates this Web site periodically to provide current information and URLs on topics relevant to all chapters.

2

Crystals of amethyst, a variety of quartz. The planar surfaces are crystal faces, which reflect the underlying arrangement of the atoms that make up the crystals. (*Chip Clark*.)

Minerals: Building Blocks of Rocks

In Chapter 1 we saw how plate tectonics describes Earth's large-scale structure and dynamics, but we touched only briefly on the wide variety of materials that appear in plate-tectonic settings. In this chapter and the next, we focus on rocks, the records of geologic processes, and on minerals, the building blocks of rocks.

To tell Earth's story accurately, geologists often adopt a Sherlock Holmesian perspective, using current evidence to deduce the processes and events that occurred in the past at some particular place on Earth. The kinds of minerals found in some volcanic rocks, for example, give evidence of the eruptions that brought molten rock, at temperatures perhaps as high as 1000°C, to Earth's surface. The minerals of a granite give evidence that it crystallized deep in the Earth's crust under conditions that formed mountains such as the Himalayas. Such conditions, which arise when two continental plates collide, produce temperatures as high as 700°C and pressures more than 10,000 times higher than at Earth's surface.

This process of deduction is essential as we attempt to understand the geology of a region and make informed guesses about the location of as yet undiscovered deposits of economically important resources such as metal ores. One of the richest sources of evidence is the focus of this chapter: **mineralogy**—the branch of geology that studies the composition, structure, appearance, stability, occurrence, and associations of minerals.

What Are Minerals?

Geologists define a **mineral** as a *naturally occurring, solid crystalline substance, generally inorganic, with a specific chemical composition*. Minerals are the building blocks of rocks, which can be made up of varying assemblages of minerals. Minerals are homogeneous: they cannot be divided by mechanical means into smaller components. With the proper tools and effort, most rocks can be separated into their constituent minerals. A few kinds of rocks, such as limestone, contain only a single kind of mineral (calcite). Others, such as granite, are made of several kinds of minerals. To identify and classify the many kinds of rocks found in the Earth and understand how they are formed, we must be informed about minerals.

We begin by examining each part of our definition of a mineral in a little more detail.

Naturally Occurring . . . To qualify as a mineral, a substance must be found in nature. Diamonds mined from the Earth in South Africa are minerals. Synthetic versions produced in industrial laboratories are not considered to be true minerals. Nor are the thousands of laboratory products invented by chemists.

Solid Crystalline Substance . . . Minerals are solid substances—they are neither liquids nor gases. When we say that a mineral is *crystalline,* we mean that the tiny particles of matter—the atoms—that compose it are arranged in an orderly, repeating, three-dimensional array. Solid materials that have no such orderly arrangement are referred to as *glassy* or *amorphous* (without form). Window glass is amorphous, as are some natural glasses formed during volcanic eruptions. Later in this chapter, we will explore in detail the process by which crystalline materials form.

Generally Inorganic . . . The stipulation that minerals are inorganic substances follows historical usage and excludes the organic substances that make up plant and animal bodies. These organic substances are made of organic carbon, the form of carbon found in all biological materials. Decaying vegetation in a swamp may be geologically transformed into coal, which is also made of organic carbon, but though it is found as a natural deposit, coal is not traditionally considered a mineral. Many minerals are, however, secreted by organisms. One such mineral, calcite, forms the shells of oysters and many other organisms, and it contains inorganic carbon. The calcite of these shells, which constitute the bulk of many limestones, fits the definition of a mineral because it is inorganic and crystalline.

. . . With a Specific Chemical Composition The key to understanding the materials and composition of the Earth lies in understanding how the chemical elements are organized into minerals. What makes each mineral unique is the combination of its chemical composition and the arrangement of its atoms in an internal structure. A mineral's chemical composition either is fixed or varies within defined limits. The mineral quartz, for example, has a fixed ratio of two atoms of oxygen to one of silicon. This ratio never varies, although quartz is found in many different kinds of rock. The components of olivine, a slightly more complex mineral, always have a fixed ratio. Iron, magnesium, and silicon make up olivine. Although the ratio of iron to magnesium atoms may vary, the sum of those atoms in relation to the number of silicon atoms always forms a fixed ratio. Figure 2.1 summarizes the definition of a mineral.

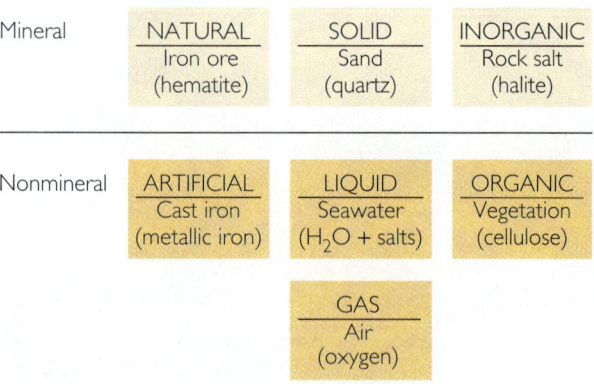

Figure 2.1 Minerals are distinguished from other materials in being naturally occurring, solid, inorganic substances with specific chemical compositions.

The Atomic Structure of Matter

A modern dictionary lists many meanings for the word *atom* and its derivatives. One of the first is "anything considered as the smallest possible unit of any material." To the ancient Greeks, *atomos* meant "indivisible." John Dalton (1766–1844), an English chemist and the father of modern atomic theory, theorized that atoms were particles of matter of several kinds that were so small that they could not be seen with any microscope and so universal that they composed all substances. In 1805 Dalton hypothesized that the various chemical elements consist of different kinds of atoms, that all atoms of any given element are identical, and that chemical compounds are formed by various combinations of atoms of different elements in definite proportions.

By the early twentieth century, physicists, chemists, and mineralogists, building on Dalton's ideas, had reached an understanding of the structure of matter, much as we know it today. We now know that an **atom** is the smallest unit of an element that retains the physical and chemical properties of that element. We also know that atoms are the small units of matter that combine in chemical reactions, but that atoms themselves are divisible into even smaller units.

The Structure of Atoms

Understanding the structure of atoms allows us to predict how chemical elements will react with one another and form new crystal structures.

The Nucleus: Protons and Neutrons At the center of every atom is a dense **nucleus** containing practically all the mass of the atom in two kinds of particles, protons and neutrons (Figure 2.2). For convenience, each of these particles is taken to have a mass of 1 atomic mass unit.[1] Protons carry an electrical charge; neutrons do not. Thus a **proton** has a positive electrical charge of +1. A **neutron** is electrically neutral—that is, uncharged. Atoms of the same chemical element may have different numbers of neutrons, but the number of their protons does not vary.

---

[1] The atomic mass unit is equal to 1/12 of the actual mass of a carbon atom with mass number 12, approximately 1.6604×10^{-24} grams.

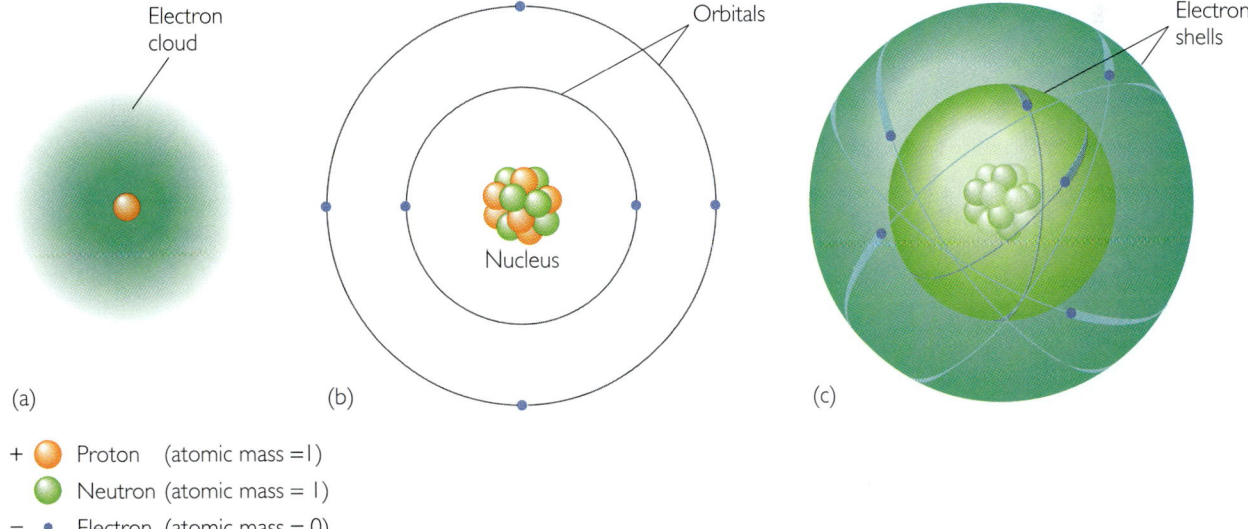

FIGURE 2.2 Electron structure of hydrogen and carbon atoms. (a) The position of the single electron of the simplest element, hydrogen, shown as an electron cloud surrounding the nucleus. (b) Electrons of a carbon atom are represented as lying in definite orbits around the nucleus, which contains six protons, each with a charge of +1, and six neutrons, each with zero charge. Six electrons are found in two concentric shells, an inner one with two electrons and an outer one with four. (c) A somewhat more realistic representation of the electron shells of a carbon atom. The size of the nucleus is greatly exaggerated; it is much too small to show on a true scale.

ELECTRONS Surrounding the nucleus is a cloud of moving **electrons,** each with a mass so small that it is conventionally taken to have no mass. Each electron carries an electrical charge of -1. The number of protons in the nucleus of any atom is balanced by the same number of electrons in the cloud outside, so an atom is electrically neutral. Modern models of atomic structure give the locations of electrons around the nucleus as *orbitals* (Figure 2.2b). They can be thought of as spherical **shells,** or regions, around the nucleus where an electron is most likely to be found, not as fixed orbits or paths of travel. For convenience in diagrams, however, we usually depict orbitals as concentric spherical shells around a nucleus (see Figure 2.2c).

Atomic Number and Atomic Mass

The number of protons in the nucleus of an atom is called its **atomic number.** Because all atoms of the same element have the same number of protons, they also have the same atomic number. All atoms with six protons, for example, are carbon atoms (atomic number 6). Since all the protons are balanced by an equal number of electrons, each element has a distinctive number of electrons, too. The atomic number of an element determines how it will react chemically with other elements.

The **atomic mass** of an element is the sum of the masses of its protons and neutrons. (Electrons, because they have so little mass, are not included in this sum.) Although the number of protons is constant, atoms of the same chemical element may have different numbers of neutrons, and therefore different atomic masses. These various kinds of atoms are called **isotopes.** Isotopes of the element carbon, for example, all with six protons, exist with six, seven, and eight neutrons, giving atomic masses of 12, 13, and 14 (see Figure 2.3). In nature, the chemical elements occur as mixtures of isotopes, so their atomic masses are never whole numbers. Carbon's atomic mass, for example, is 12.011. It is close to 12 because the isotope carbon-12 is overwhelmingly abundant. The relative abundance of the different isotopes of an element on Earth is determined by geological processes that enhance the abundance of some isotopes over others. Carbon-12, for example, is favored by some reactions, such as photosynthesis, in which organic carbon compounds are produced from inorganic carbon compounds.

CHEMICAL REACTIONS

The structure of any particular kind of atom determines its chemical reactions with other atoms. **Chemical reactions** are interactions of the atoms of two or more chemical elements in certain fixed proportions that produce new chemical substances—

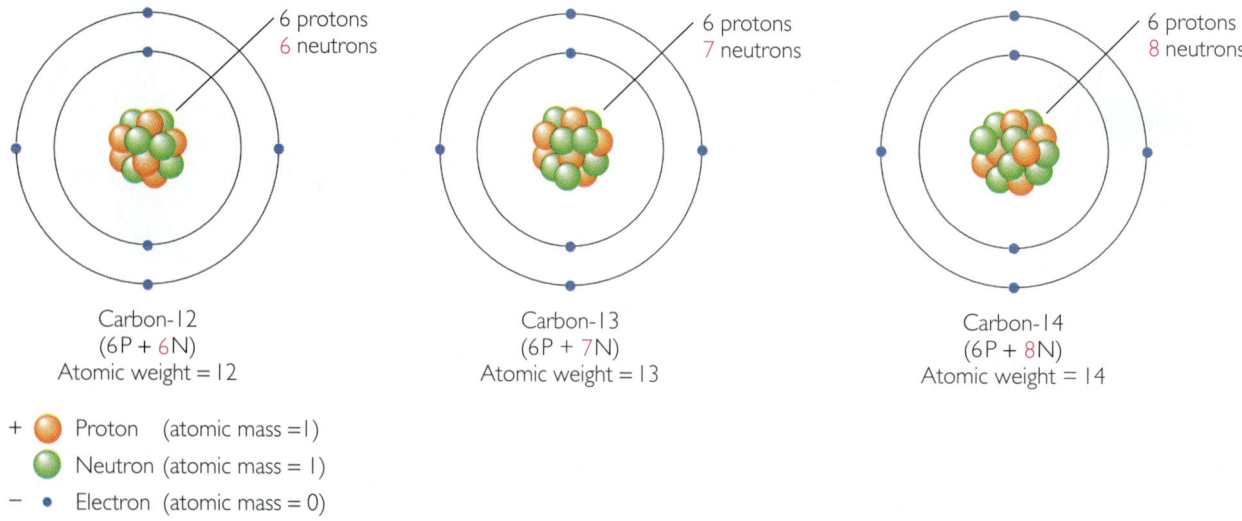

FIGURE 2.3 These three carbon isotopes all have the same number of protons and thus the same atomic number, 6. Their atomic masses differ, however, because they have slightly different numbers of neutrons. The atomic mass of any element is the average of the weighted sum of the atomic masses of its various isotopes. One isotope of an element—for example, carbon-12—is far more abundant than the others because natural processes favor that particular isotope.

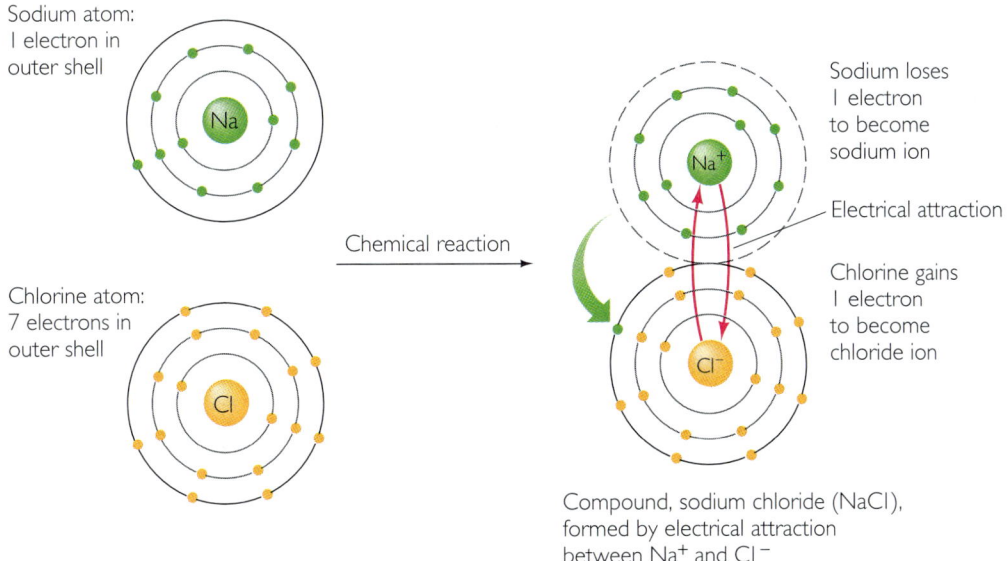

FIGURE 2.4 Formation of a chemical compound by reaction of chlorine and sodium atoms. When sodium (Na) and chlorine (Cl) react to form table salt (sodium chloride, NaCl), the sodium atom loses one electron from its outer shell and the chlorine atom acquires that electron. During this reaction, an orderly array of ions is formed.

chemical compounds. When two hydrogen atoms combine with one oxygen atom, they form a new chemical compound that we call water (H_2O). The properties of a chemical compound formed in the course of a reaction may be entirely different from those of its constituent elements. For example, when an atom of sodium, a metal, combines with an atom of chlorine, a noxious gas, they form the chemical compound sodium chloride, better known as table salt. We represent this compound by the chemical formula *NaCl,* the symbol *Na* standing for the element sodium and the symbol *Cl* for the element chlorine. (Every chemical element has been assigned its own symbol, which we use as a kind of shorthand for writing chemical formulas and equations.)

Chemical reactions take place primarily through the interactions of electrons. To understand those interactions we need to know the number of electrons in an atom and how they are arranged in electron shells.

Gaining and Losing Electrons

Electrons surround the nucleus of an atom in a unique set of concentric spheres called electron shells, each of which can hold up to a certain number of electrons. In the chemical reactions of most elements, only the electrons in the outermost shells interact. In the reaction between sodium (Na) and chlorine (Cl) that forms sodium chloride (NaCl), the sodium atom loses an electron from its outer shell of electrons and the chlorine atom gains an electron in its outer shell (Figure 2.4).

IONS After *gain or loss of an electron,* the atoms in the new chemical compound are no longer electrically neutral. When the sodium atom loses an electron, it becomes a sodium **ion,** with an electrical charge of +1, because it still has the same number of protons but one less electron. It is now represented by the symbol Na^+. The chlorine atom, in gaining an electron, has become a *chloride* ion with an electrical charge of −1, written as Cl^-. Positive ions, such as sodium, are called **cations;** negative ions, such as chloride, are called **anions.** The compound NaCl itself remains electrically neutral because the positive charge on Na^+ is exactly balanced by the negative charge on Cl^-.

Groups of ions may join to form *complex ions,* such as the common sulfate ion (SO_4^{2-}), a component of the mineral anhydrite ($CaSO_4$) and an abundant constituent of seawater. The sulfate ion is a unit made up of one sulfur ion with a +6 charge and four oxygen ions, each with a −2 charge, the net charge adding to −2.

ELECTRON SHELLS AND ION STABILITY Before reacting with chlorine, the sodium atom has one electron in its outer shell. When it loses that electron, its outer shell is eliminated and the next shell inward, which has eight electrons (the maximum this shell can hold), becomes the outer shell. The original chlorine atom had seven electrons in its outer shell, with room for a total of eight. By gaining an electron, its outer shell is filled. Many elements have a strong tendency to acquire a full outer electron shell, some by gaining electrons and some by losing them in the course of a chemical reaction. The stability of ions with fully occupied outer shells is related to the interactions of electrons in various orbitals around the nucleus.

Many chemical reactions involve gains and losses of several electrons as two or more elements combine. The element calcium (Ca), for example, becomes a doubly charged cation, Ca^{2+}, as it reacts with two chlorine atoms to form calcium chloride. (In the chemical formula for calcium chloride, $CaCl_2$, the presence of two chloride ions is symbolized by the subscript 2. Chemical formulas thus show the relative proportion of atoms or ions in a compound. Common practice is to omit the subscript 1 next to single ions in a formula. For the single Ca in $CaCl_2$, therefore, the subscript 1 is omitted.)

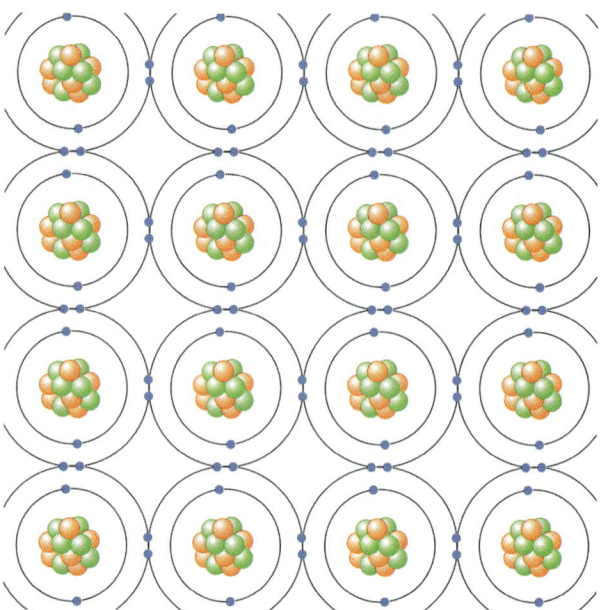

FIGURE 2.5 Electron sharing in diamond. The mineral diamond is composed of the single element carbon. Each carbon atom has four electrons in its outer shell, and each acquires four more by sharing its electrons with four adjacent carbon atoms.

Electron Sharing

Not all chemical elements react by gaining or losing electrons. Some have a strong tendency to combine chemically by engaging in **electron sharing** with atoms of the same or a different element to achieve a stable configuration of electrons. Carbon and silicon, two of the abundant elements of Earth's crust, are elements that tend toward electron sharing.

The mineral diamond is composed of the single element carbon. Each carbon atom has four electrons in its outer shell, and each acquires four more by sharing its electrons with four adjacent carbon atoms (see Figure 2.5). When these electrons are shared, all of the atoms act as if each had a full complement of eight electrons in its outer shell. The shared electrons cannot be considered to have been gained or lost. In a sense, the nuclei of the sharing atoms have "gained" the electron for whatever part of the time it can be visualized as belonging to the outer shell of one or the other atom. Nevertheless, because the atoms still have their original number of electrons, we do not ordinarily refer to them as ions.

Periodic Table of Elements

Chemists have long known that some groups of elements have similar chemical properties, such as boiling and melting points and tendencies to react chemically with other elements. These groups differ markedly from one another. As the atomic structure of elements became known, these properties proved to correspond to the electron shell patterns of the elements.

The periodic table (Figure 2.6) organizes the elements (from left to right along a row) in order of atomic number (number of protons), which also means increasing numbers of electrons in the outer shell. The third row from the top, for example, starts at the left with sodium (atomic number 11), which has one electron in its outer shell. The next is magnesium (atomic number 12), which has two electrons in its outer shell, followed by aluminum (atomic number 13), with three, and silicon (atomic number 14), with four. Then come phosphorus (atomic number 15), with five; sulfur (atomic number 16), with six; and chlorine (atomic number 17), with seven. The last element in this row is argon (atomic number 18), with eight electrons, the maximum possible, in its outer shell. Each column in the table forms a vertical grouping of elements with similar electron shell patterns.

ELEMENTS THAT TEND TO LOSE ELECTRONS

The elements in the leftmost column all have a single electron in their outer shells and have a strong tendency to lose that electron in chemical reactions.

Of this group, hydrogen (H), sodium (Na), and potassium (K) are found in major abundance at Earth's surface and in its crust.

The second column from the left includes two more elements of major abundance, magnesium (Mg) and calcium (Ca). Elements in this column have two electrons in their outer shells and a strong tendency to lose them both in chemical reactions.

ELEMENTS THAT TEND TO GAIN ELECTRONS Toward the right side of the table, the two columns headed by oxygen (O), the most abundant element in the Earth, and fluorine (F), a highly reactive toxic gas, group the elements that tend to gain electrons in their outer shells. The elements in the column headed by oxygen have six of the possible eight electrons in their outer shells and tend to gain two electrons. Those in the column headed by fluorine have seven electrons in their outer shells and tend to gain one.

OTHER ELEMENTS The columns between the two on the left and the two headed by oxygen and fluorine have varying tendencies to gain, lose, or share electrons. The column toward the right side of the table headed by carbon (C) includes silicon (Si), of major abundance in the Earth. As we noted earlier, both silicon and carbon tend to share electrons.

The elements in the last column on the right, headed by helium (He), have full outer shells and thus no tendency either to gain or to lose electrons. As a result, these elements, in contrast to those in other columns, do not react chemically with other elements, except under very special conditions.

We can predict a great many chemical reactions from the information conveyed by the columns and rows in the periodic table. Figure 2.7 illustrates diverse patterns of gaining, losing, or sharing electrons in a comparison of five common elements: hydrogen, sodium, magnesium, oxygen, and chlorine.

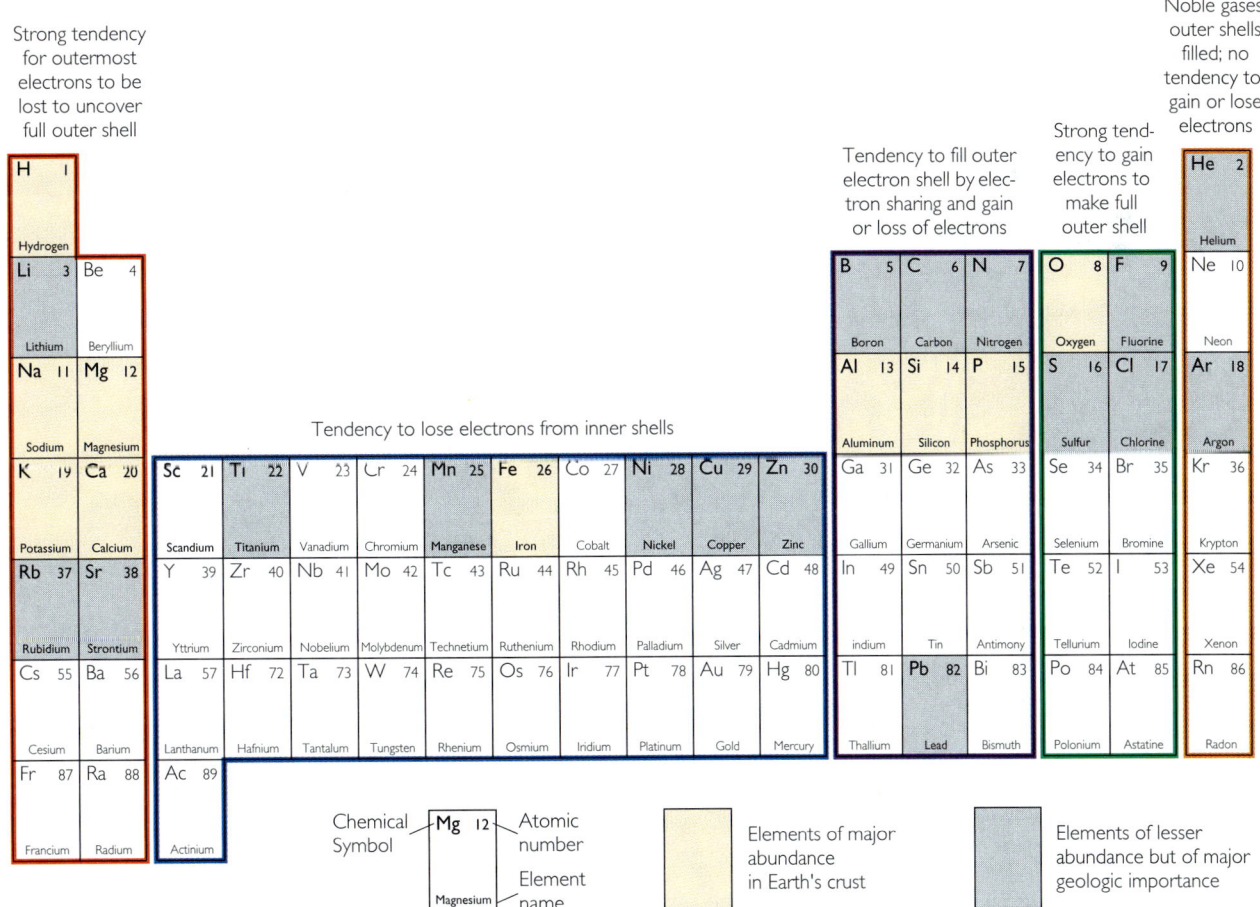

FIGURE 2.6 Periodic table of elements. Characteristics associated with different electron shell patterns are indicated at the top of the table. (Two special groups of rare elements—atomic numbers 58–71 and 90–103—have been omitted.)

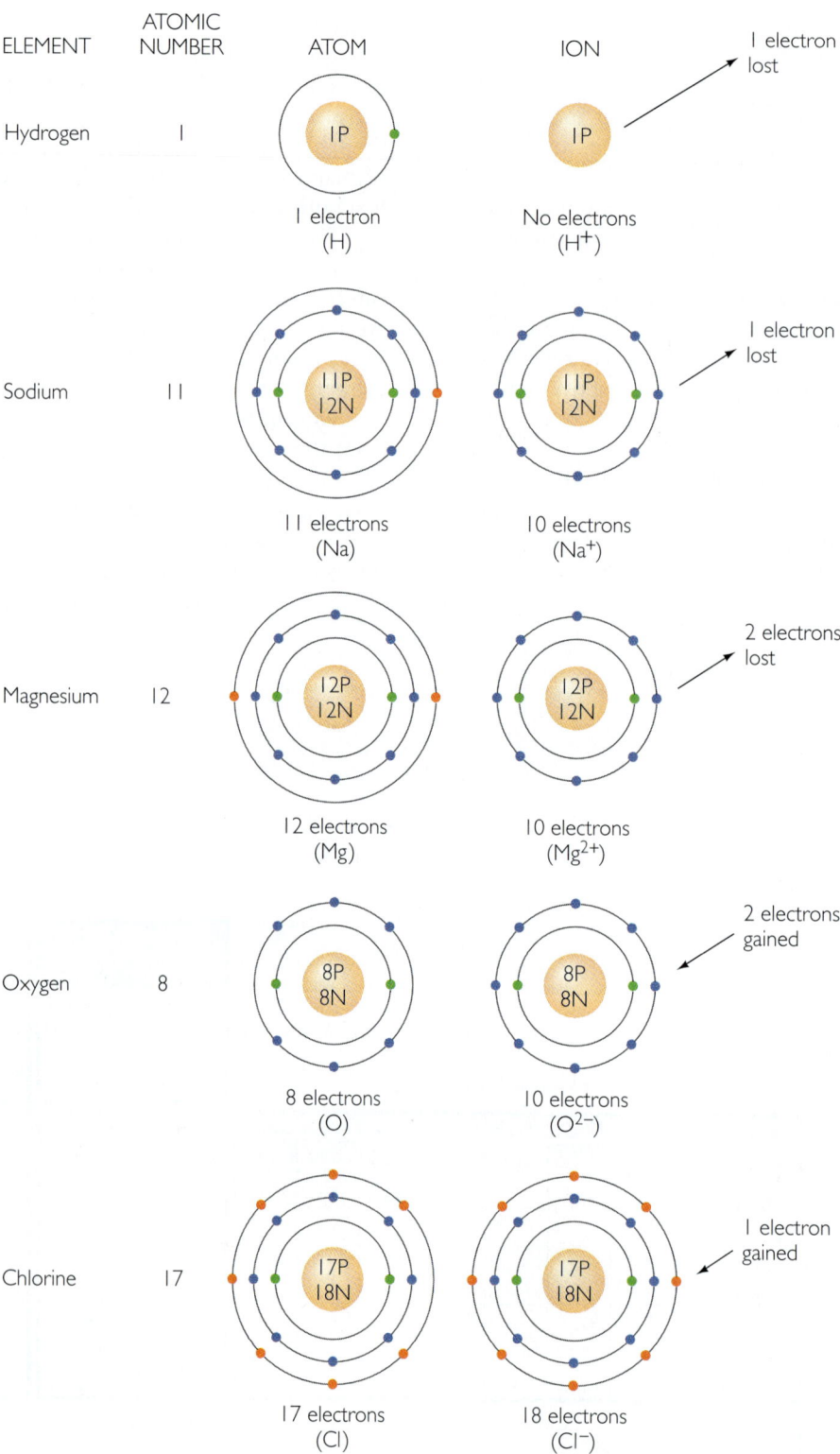

FIGURE 2.7 These models of the atoms and ions of five common elements illustrate the diversity of electron shells. Ions are formed as electrons are gained or lost from outer shells. (P stands for proton, N for neutron.)

CHEMICAL BONDS

The ions or atoms of elements that make up compounds are held together by electrical forces of attraction between electrons and protons, which we call chemical bonds. The electrical attractions either of shared electrons or of gained or lost electrons may be strong or weak, and the bonds created by these attractions are correspondingly strong or weak. Strong chemical bonds keep a substance from chemically decomposing into its elements or into other compounds. They also make minerals hard and keep them from cracking or splitting. Two major types of bonds are found in most rock-forming minerals: ionic bonds and covalent bonds.

Ionic Bonds

The simplest form of chemical bond is the **ionic bond.** Bonds of this type are formed by electrical attraction between ions of the opposite charge, such as Na^+ and Cl^- in sodium chloride (see Figure 2.4). This attraction is of exactly the same nature as the static electricity that can make clothing of nylon or silk cling to the body. The strength of an ionic bond decreases greatly as the distance between ions increases. Bond strength increases as the electrical charges of the ions increase. Ionic bonds are the dominant type of chemical bonds in mineral structures; *about 90 percent of all minerals are essentially ionic compounds.*

Covalent Bonds

Elements that do not readily gain or lose electrons to form ions and instead form compounds by sharing electrons are held together by **covalent bonds.**

Covalent bonds are generally stronger than ionic bonds. One mineral with a covalently bound crystal structure is diamond, consisting of the single element carbon. As we saw in the case of diamond, carbon has four electrons in its outer shell and acquires four more by electron sharing to achieve a full outer shell with eight electrons. In diamond, every carbon atom (not an ion) is surrounded by four others arranged in a regular *tetrahedron,* a four-sided pyramidal form, each side a triangle (Figure 2.8a). In this configuration, each carbon atom shares an electron with each of its four neighbors and thus achieves a stable set of eight electrons in its outer shell.

Atoms of metallic elements, which have strong tendencies to lose electrons, pack together as cations, while freely mobile electrons are shared and dispersed among the ions. This free electron sharing results in a kind of covalent bond that we call a **metallic bond.** It is found in a small number of minerals, among them the metal copper and some sulfides.

The chemical bonds of some minerals are intermediate between pure ionic and pure covalent bonds because some electrons are exchanged and others are shared.

ATOMIC STRUCTURE OF MINERALS

Minerals can be looked at in two complementary ways: as crystals (or grains) we can see with the naked eye, and as assemblages of submicroscopic atoms organized in an ordered three-dimensional array (Figure 2.8). The preceding discussion of how substances are formed by chemical bonding between atoms and ions prepares us for a closer view of the

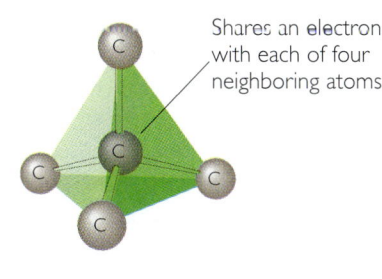

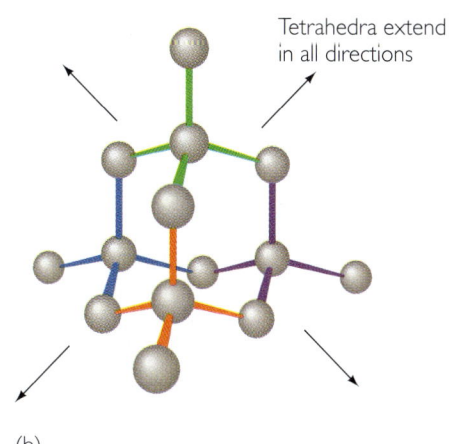

FIGURE 2.8 The carbon tetrahedron of diamond. (a) A single tetrahedron formed by a carbon atom bonded to four other carbon atoms. (b) A network of carbon tetrahedra linked to one another.

orderly forms that characterize minerals' structure and of the conditions under which minerals form. Later in this chapter, you will see that the crystal structures of minerals are reflected in their physical properties. First, however, we turn to the question of how minerals form.

How Do Minerals Form?

Minerals are formed by the process of **crystallization,** the growth of a solid from a material whose constituent atoms can come together in the proper chemical proportions and crystalline arrangement. (Remember that the atoms in a mineral are arranged in an ordered, three-dimensional array.) The bonding of carbon atoms in diamond, a covalently bonded mineral, is one example of crystallization and crystal structure. In satisfying the requirements of electron sharing, carbon atoms come together in tetrahedra, each tetrahedron attaching to another and building up a regular three-dimensional structure from a great many atoms (see Figure 2.8b). As a diamond crystal grows, it extends its tetrahedral structure in all directions, always adding new atoms in the proper geometric arrangement. Diamonds can be synthesized under very high pressures and temperatures that mimic conditions in Earth's mantle.

The sodium and chloride ions that make up sodium chloride, an ionically bonded mineral, also crystallize in an orderly three-dimensional array. In Figure 2.9 we can see the geometry of their arrangement, with each ion of one kind surrounded by six ions of the other in a series of cubes extend-

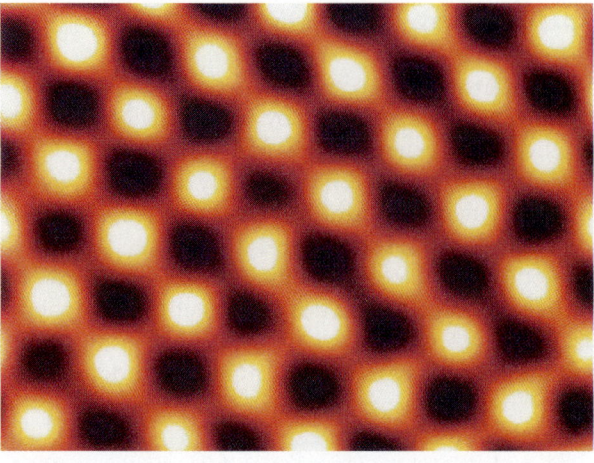

(a)

(b)

FIGURE 2.10 (a) Ultra-high vacuum scanning tunneling microscope image of a galena (PbS) surface with atomic resolution. The scale of the image is approximately $26 \times 10^{-8} \times 26 \times 10^{-8}$ cm. The cubic structure of galena is reflected in the arrangement of both the bright and dark spots. Theoretical calculations suggest that the bright spots correspond to lead atoms and the dark spots correspond to sulfur sites. *(Kevin M. Rosso and Michael F. Hochella, Jr., Virginia Polytechnic and State University.)* (b) Galena crystals. *(Chip Clark.)*

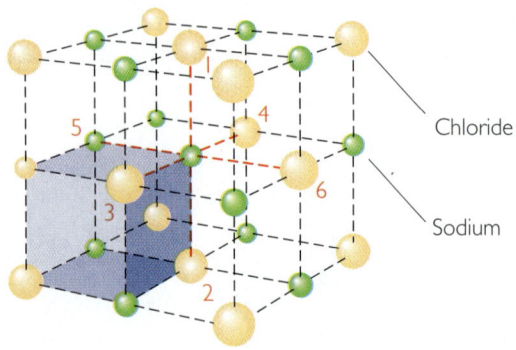

FIGURE 2.9 Structure of halite (sodium chloride). The dotted lines between the ions show the cubic geometry of this mineral; they do not represent bonds. In this example it is easy to see one sodium ion surrounded by six chloride ions. (Ions are not drawn to scale.)

ing in three directions. This arrangement is called a cubic structure.

Atoms and ions are so small—most of them only a few ten-millionths of a centimeter—that we cannot see the crystalline arrangement of a mineral directly even with the most powerful ordinary microscope. With especially high-powered electron and other, newer kinds of microscopes, however, we can now image the atomic arrangements of crystals (Figure 2.10).

Crystallization starts with the formation of microscopic single **crystals,** bodies whose boundaries are natural flat (plane) surfaces. These surfaces, called *crystal faces,* are the defining external characteristic of a crystal. The crystal faces of a mineral are the external expression of the mineral's internal atomic structure. Figure 2.11 pairs drawings of perfect crystals (which are very rare in nature) and photos of two actual minerals. The simple geometric cubes of sodium chloride crystals (the mineral halite, or rock salt) correspond to the cubic arrangement of its ions. The six-sided (hexagonal) shape of the quartz crystal corresponds to its hexagonal internal atomic structure.

During crystallization, the initially microscopic crystals grow larger, maintaining their crystal faces as long as they are free to grow. Large crystals with well-defined faces form when growth is slow and steady and space is adequate to allow growth without interference from other crystals nearby. For this reason, most large mineral crystals form in open spaces in rocks, such as open fractures or cavities (Figure 2.12).

Often, however, the spaces between growing crystals fill in or crystallization proceeds too rapidly. Crystal faces then grow over one another, and the former crystals coalesce to become a solid mass of crystalline particles, or *grains*. In the crystallized mass, few or no grains show crystal faces (see Figure 2.12). Large crystals that can be seen with the naked eye are relatively unusual, but many microscopic minerals in rocks display crystal faces.

Glassy materials, which solidify from liquids so quickly that they lack any internal atomic order, do not form crystals with plane faces. Instead they are found as masses with curved, irregular surfaces. The most common glass is volcanic glass.

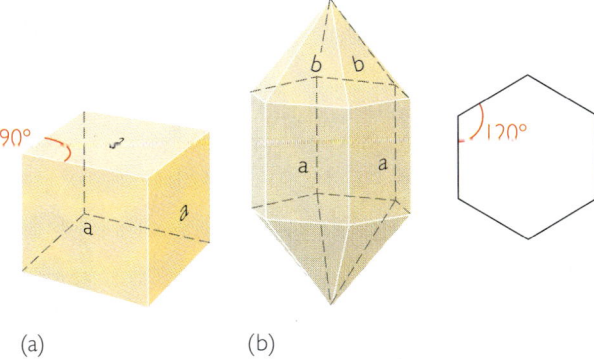

FIGURE 2.11 Perfect crystals. A perfect crystal is rare, but no matter how irregular the shapes of the faces may be, the angles are always exactly the same. (a) Halite, a cubic crystal: structural diagram and sample of crystals *(Ed Degginger/Bruce Coleman.)* (b) Quartz, a hexagonal crystal: structural diagrams and sample of a nearly perfect crystal. *(Breck P. Kent.)* The cross section at right angles to the long dimension of the crystal shows a regular hexagon with "A" faces at 120° angles.

FIGURE 2.12 This sample of quartz crystals displays both macro- and microcrystallinity. Large, well-formed crystals were able to form in the central cavity where adequate space could accommodate their slow growth. The exterior of this mass is rimmed by a whitish layer of smaller crystal grains displaying few crystal faces, as would be expected in a space confined by other, similar masses of crystals. *(Chip Clark.)*

When Do Minerals Form?

Lowering the temperature of a liquid below its freezing point is one way to start the process of crystallization. In the case of water, 0°C is the temperature below which crystals of ice, a mineral, start to form. Similarly, magma, a hot, molten liquid rock, crystallizes solid minerals when it cools. As a magma falls below its melting point, which may be over 1000°C, crystals of silicate minerals such as olivine or feldspar begin to form. (Geologists usually refer to melting points of magmas rather than freezing points, since freezing normally implies cold.)

Another set of conditions that can produce crystallization occurs during precipitation, as liquids evaporate from a solution. A solution is formed when one chemical substance is dissolved in another, such as salt in water. As the water evaporates from a salt solution, the concentration of salt eventually gets so high that the solution is said to be saturated—it can hold no more salt. If evaporation continues, the salt starts to **precipitate,** or drop out of solution as crystals. Deposits of halite or table salt form under just these conditions when seawater evaporates to the point of saturation in some hot, arid bays or arms of the ocean.

Crystals also form when atoms and ions in solids become mobile and rearrange themselves at high temperatures. For most minerals, temperatures must reach at least 250°C before this kind of rearrangement forms new minerals with different crystal structures. The mineral mica forms this way.

Two major factors control the arrangement of atoms and ions in a crystal structure: the number of neighboring atoms or ions, and their size.

Sizes of Ions

We can think of ions as if they were solid spheres, packed together in close-fitting structural units. Figure 2.13 shows the relative sizes of the ions in NaCl. There are six neighboring ions in NaCl's basic structural unit. The relative sizes of the sodium (smaller) and chloride ions allow them to fit together in a closely packed arrangement.

Ion size is related to the atomic structures of the elements (Figure 2.14). The sizes of ions increase with the number of electrons and electron shells. An ion's charge also affects its size. The more electrons an element loses to become a cation, the stronger its positive charge and the stronger the electrical attraction of its nucleus for the remaining electrons. Many of the cations of abundant minerals are relatively small; most anions are large. This is the case with the most common Earth anion, oxygen. Because anions

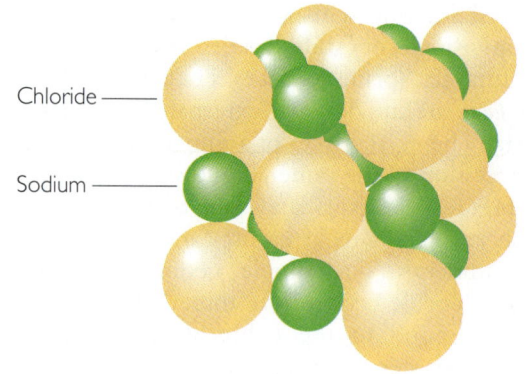

FIGURE 2.13 The relative sizes of sodium and chloride ions allow them to pack together in a cubic structure. Ions here are shown in their correct relative sizes.

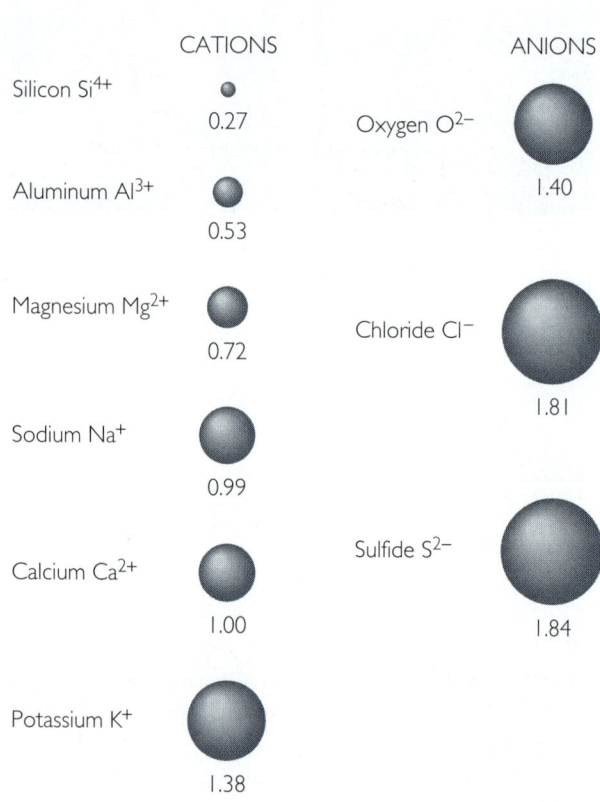

FIGURE 2.14 Sizes of ions as they are commonly found in rock-forming minerals. Ionic radii are given in 10^{-8} cm. (After L. G. Berry, B. Mason, and R. V. Dietrich, *Mineralogy* [San Francisco: W. H. Freeman, 1983].)

tend to be larger than cations, it is apparent that most of the space of a crystal is occupied by the anions and that cations fit into the spaces between them. As a result, crystal structures are determined largely by the way the anions are arranged and the way the cations fit between them.

CATION SUBSTITUTION: SAME CRYSTAL STRUCTURE, DIFFERENT CHEMICAL COMPOSITIONS Cations of similar sizes and charges tend to substitute for one another and to form compounds with the same crystal structure but differing chemical composition. Cation substitution is common in silicate minerals, in which cations combine with the silicate ion $(SiO_4)^{4-}$. This process is illustrated by olivine, a mineral abundant in many volcanic rocks.

Iron (Fe) and magnesium (Mg) ions are similar in size, and both have two positive charges, so they easily substitute for each other in the structure of olivine. The composition of pure magnesium olivine is Mg_2SiO_4; the pure iron olivine is Fe_2SiO_4. The composition of olivine with both iron and magnesium is given by the formula $(Mg,Fe)_2SiO_4$, which simply means that the number of iron and magnesium cations may vary, but their combined total (expressed as a subscript 2) does not vary in relation to each $(SiO_4)^{4-}$ ion. The proportion of iron to magnesium is determined by the relative abundance of the two elements in the molten material from which the olivine crystallized. In many silicate minerals, aluminum (Al) substitutes for silicon (Si). Aluminum and silicon ions are so similar in size that aluminum can take the place of silicon in many crystal structures. The difference in charge between aluminum (3+) and silicon (4+) ions is balanced by an increase in one of the other cations, such as sodium (1+).

POLYMORPHS: DIFFERENT CRYSTAL STRUCTURE, SAME CHEMICAL COMPOSITION The same combinations of elements in the same proportions can sometimes form more than one kind of crystal structure, and therefore more than one kind of mineral. These alternative possible structures for a single chemical compound are called **polymorphs** ("many forms"). The structure that actually forms depends on the conditions of pressure and temperature, and therefore on the depth within the Earth at the time and place crystallization occurs.

Diamond and graphite (the material that is used as the "lead" in pencils) are polymorphs. These two minerals, both formed from carbon, have different crystal structures and very different appearances (Figure 2.15). From experimentation and geological observation, we know that diamond forms and remains stable at the very high pressures and temperatures of Earth's mantle. The high pressure in the mantle forces the atoms in diamond to be closely packed. Diamond therefore has a higher density (mass per unit volume), 3.5 g/cm³, than graphite, which is less closely packed and has a density of only 2.1 g/cm³. Graphite forms and is stable at relatively moderate pressures and temperatures, such as those in Earth's crust.

Low temperatures can also produce closer packing. Quartz and cristobalite are polymorphs of silica (SiO_2). Quartz is formed at low temperatures and is relatively dense (2.7 g/cm³). Cristobalite, formed at a higher temperature, has a more open structure and is therefore less dense (2.3 g/cm³).

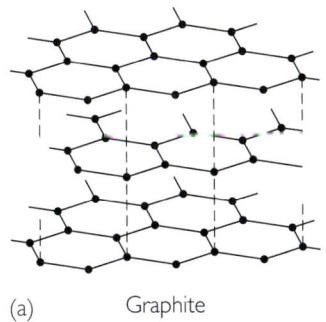
(a) Graphite

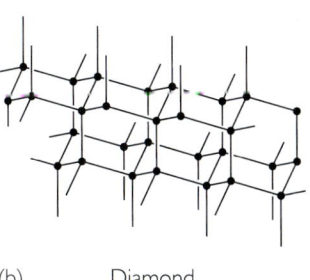
(b) Diamond

FIGURE 2.15 (a) In graphite, sheets of carbon atoms arranged in hexagons are stacked above one another with weak bonds (dashed lines) between the sheets. (b) In diamond, carbon atoms are arranged in a tetrahedral network. Graphite (at the left in photo) forms flat, platelike masses of crystals. Diamond (at the right in photo), when well crystallized, typically forms octahedral (eight-faced) crystals. *(Chip Clark.)*

Rock-Forming Minerals

All minerals have been grouped into eight classes according to their chemical composition; six of those classes of minerals are shown in Table 2.1. Some minerals, such as copper, occur naturally as un-ionized pure elements, and they are classified as *native elements*. Most others are classified by their anions. Olivine, for example, is classed as a silicate by its silicate anion, $(SiO_4)^{4-}$. Another mineral we have discussed, halite (NaCl), is classed as a halide from its chloride anion, Cl^-. So is its close relative sylvite, potassium chloride (KCl). The anion of the carbonate minerals is $(CO_3)^{2-}$.

Although many thousands of minerals are known, geologists commonly encounter only about thirty of them. These are the minerals that are the building blocks of most crustal rocks, and thus they are called *rock-forming minerals*. Their relatively small number reflects the small number of elements that are found in major abundance in Earth's crust: as we saw in Chapter 1, 99 percent of Earth's crust is made up of only nine elements. Figure 1.7 shows the relative abundance, by weight, of these elements.

In the following pages, we discuss the most common rock-forming minerals:

- *Silicates,* which are the most abundant minerals in Earth's crust and are composed of oxygen (O) and silicon (Si)—the two most abundant elements in Earth's crust—mostly in combination with the cations of other elements.
- *Carbonates,* minerals made of carbon and oxygen in the form of the carbonate anion $(CO_3)^{2-}$, in combination with calcium and magnesium; calcite $(CaCO_3)$ is one such mineral.
- *Oxides,* a group of compounds of oxygen and metallic cations, such as the mineral hematite (Fe_2O_3).
- *Sulfides,* compounds of the sulfide anion S^{2-}, and metallic cations, a group that includes the mineral pyrite (FeS_2).
- *Sulfates,* compounds of the sulfate anion $(SO_4)^-$ and metallic cations and including the mineral anhydrite $(CaSO_4)$.

The other chemical classes of minerals, including native elements and halides, are not found as commonly as rock-forming minerals.

Silicates

The basic building block of all silicate mineral structures is the *silicate ion,* formed by four oxygen ions (O^{2-}) surrounding and sharing electrons with a silicon ion (Si^{4+}), giving the formula $(SiO_4)^{4-}$ (Figure 2.16a). This configuration results in a four-sided pyramidal form, called a *tetrahedron,* in which each side forms a triangle (Figure 2.16b). Each silicon-oxygen tetrahedron is an anion with four negative charges, which must be balanced by four positive charges to make an electrically neutral mineral in one of two ways: The ion can bond with cations such as sodium (Na^+), potassium (K^+), calcium (Ca^2), magnesium (Mg^{2+}), and iron (Fe^{2+}). Alternatively, the ion can share oxygens with other silicon-oxygen tetrahedra. All silicate minerals are made up

TABLE 2.1
Some Chemical Classes of Minerals

CLASS	DEFINING ANIONS	EXAMPLE
Native elements	None: no charged ions	Copper metal (Cu)
Oxides and hydroxides	Oxygen ion (O^{2-}) Hydroxyl ion (OH^-)	Hematite (Fe_2O_3) Brucite $(Mg[OH]_2)$
Halides	Chloride (Cl^-), fluoride (F^-), bromide (Br^-), iodide (I^-)	Halite (NaCl)
Carbonates	Carbonate ion (CO_3^{2-})	Calcite $(CaCO_3)$
Sulfates	Sulfate ion (SO_4^{2-})	Anhydrite $(CaSO_4)$
Silicates	Silicate ion (SiO_4^{4-})	Olivine (Mg_2SiO_4)

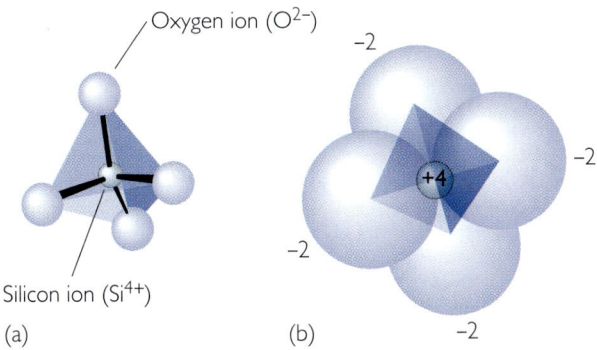

FIGURE 2.16 Silicate ion. (a) A model showing the structure of the silicon-oxygen tetrahedron. In the center of this figure is one silicon ion carrying a positive charge of 4. It is surrounded by four oxygen ions, each carrying a minus charge of 2. After bonding, the resulting silicate ion carries a negative charge of 4 (4 minus 8 equals -4). (b) A more realistic model of the silicate ion with the atoms drawn to scale and filling space.

of silicon-oxygen tetrahedra as basic units, linked in combinations of these two ways. Tetrahedra may be isolated, or they may be linked in rings, single chains, double chains, sheets, or frameworks, as shown in Figure 2.17.

ISOLATED TETRAHEDRA Isolated tetrahedra are linked by the bonding of each oxygen ion of the tetrahedron to a cation (Table 2.2; Figure 2.17a); the cations in turn bond to the oxygens of other tetrahedra. The tetrahedra are thus isolated from one another by cations on all sides. Olivine is a rock-forming mineral with this structure.

RING LINKAGES Rings of tetrahedra form as two oxygens of each tetrahedron bond to adjacent tetrahedra in closed rings (Table 2.2; Figure 2.17b). In these rings, each tetrahedron shares two of its oxygens with other tetrahedra, one on each side. Rings may link three, four, or six tetrahedra. Cordierite, common in metamorphic rocks, is a rock-forming mineral with this structure.

SINGLE-CHAIN LINKAGES Single chains also form by sharing oxygens. Two oxygens of each tetrahedron bond to adjacent tetrahedra, but in an open-ended chain instead of a closed ring (Figure 2.17c). Single chains are linked to other chains by cations. Minerals of the pyroxene group are single-chain silicate minerals. Enstatite, a pyroxene, is composed of iron and/or magnesium ions limited to a chain of tetrahedra in which the two cations may substitute for each other, as in olivine. The formula $(Mg,Fe)SiO_3$ reflects this structure.

DOUBLE-CHAIN LINKAGES Two single chains may combine to form double chains linked to each other by shared oxygens (Table 2.2; Figure 2.17d). Adjacent double chains linked by cations form the structure of the amphibole group of minerals. Hornblende, a member of this group, is an extremely common mineral in both igneous and metamorphic rocks. It has a complex composition including calcium (Ca^{2+}), sodium (Na^+), magnesium (Mg^{2+}), iron (Fe^{2+}), and aluminum (Al^{3+}).

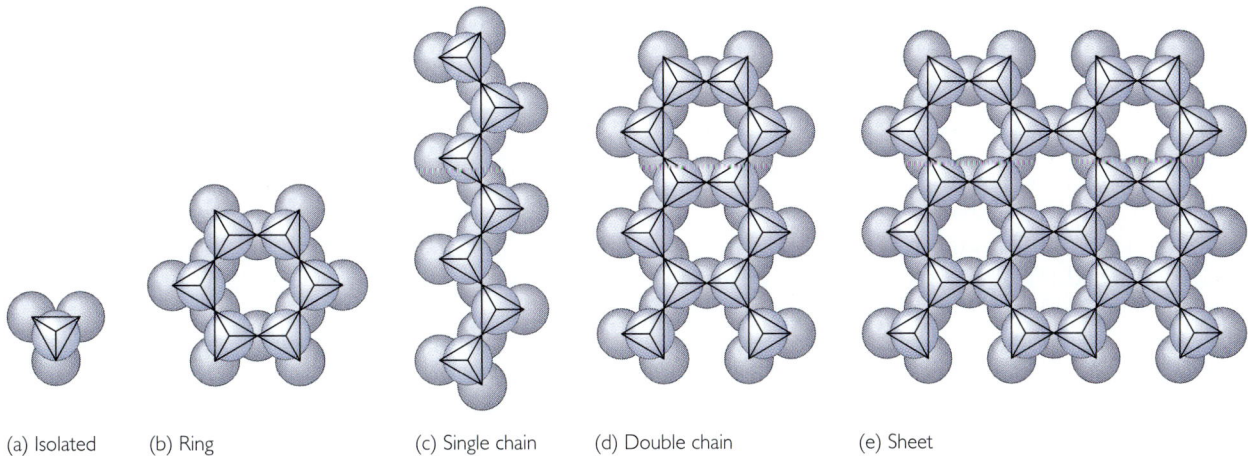

(a) Isolated   (b) Ring   (c) Single chain   (d) Double chain   (e) Sheet

FIGURE 2.17 The crystal structures of silicate minerals, which are classified according to the different ways in which silica tetrahedra can be linked.

TABLE 2.2

MAJOR SILICATE STRUCTURES

GEOMETRY OF LINKAGE OF SiO$_4$ TETRAHEDRA	EXAMPLE MINERAL	CHEMICAL COMPOSITION
Isolated tetrahedra: No sharing of oxygens between tetrahedra; individual tetrahedra linked to each other by bonding to cation between them	Olivine	Magnesium-iron silicate (Mg,Fe)$_2$SiO$_4$
Rings of tetrahedra: Joined by shared oxygens in three-, four-, or six-membered rings	Cordierite	Magnesium-iron-aluminum silicate Al$_3$(Mg,Fe)$_2$Si$_5$AlO$_{18}$
Single chains: Each tetrahedron linked to two others by shared oxygens; chains bonded by cations	Pyroxene (enstatite)	Magnesium-iron silicate (Fe,Mg)SiO$_3$
Double chains: Two parallel chains joined by shared oxygens between every other pair of tetrahedra; to cations that lie between the chains	Amphibole (hornblende)	Calcium-magnesium-iron silicate Ca(Mg,Fe)$_4$Al(Si$_7$Al)O$_{22}$(OH,F)
Sheets: Each tetrahedron linked to three others by shared oxygens; sheets bonded by cations	Kaolinite Mica (muscovite)	Aluminum silicate Al$_2$Si$_2$O$_5$(OH)$_4$ Potassium-aluminum silicate KAl$_3$Si$_3$O$_{10}$(OH)$_2$
Frameworks: Each tetrahedron shares all its oxygens with other SiO$_4$ tetrahedra (in quartz) or AlO$_4$ tetrahedra	Feldspar (orthoclase) Quartz	Potassium-aluminum silicate KAlSi$_3$O$_8$ Silicon dioxide SiO$_2$

SHEET LINKAGES Sheets are structures in which each tetrahedron shares three of its oxygens with adjacent tetrahedra to build stacked sheets of tetrahedra (Table 2.2; Figure 2.17e). Cations may be interlayered with tetrahedral sheets. The micas (Figure 2.18) and clay minerals are the most abundant sheet silicates. Muscovite (KAl$_3$Si$_3$O$_{10}$[OH]$_2$) is one of the commonest of sheet silicates, found in many types of rocks. It can be separated into extremely thin, transparent sheets. Kaolinite (Al$_2$Si$_2$O$_5$[OH]$_4$), which also has this structure, is a common clay mineral found in sediments and is the basic raw material for pottery and china.

FRAMEWORKS Three-dimensional frameworks form as each tetrahedron shares all its oxygens with other tetrahedra (see Table 2.2). Members of the feldspar group, the most abundant minerals in Earth's crust, are framework silicates, as is another of the most common minerals, quartz (SiO$_2$) (Figure 2.18).

FIGURE 2.18 Silicate minerals (clockwise from upper left): feldspar, a framework structure; mica, a sheet structure; pyroxene, a single-chain structure; quartz, a framework structure; and olivine, an isolated tetrahedral structure. *(Chip Clark.)*

COMPOSITION OF SILICATES Chemically, the simplest silicate is silicon dioxide, also called silica (SiO_2), which is found most frequently as the mineral quartz. The tendency of silicon to bond with oxygen is so strong that silicon is never found in nature as the pure element; it is always combined with oxygen. (The pure silicon used in computer chips is artificially prepared by advanced chemical techniques.) When the silicate tetrahedra of quartz are linked, sharing two O's for each Si, the total formula adds up to SiO_2.

In other silicate minerals, the basic units—rings, chains, sheets, and frameworks—are bonded to such cations as sodium (Na^+), potassium (K^+), calcium (Ca^{2+}), magnesium (Mg^{2+}), and iron (Fe^{2+}). As we noted in our discussion of cation substitution, aluminum (Al^{3+}) substitutes for silicon in many silicate minerals. All of the several varieties of feldspar, for example, contain aluminum, with various combinations of potassium, sodium, and calcium.

Carbonates

Of the nonsilicate minerals (Figure 2.19), the mineral calcite (calcium carbonate, $CaCO_3$) is one of

FIGURE 2.19 Nonsilicate minerals (clockwise from upper left): halite, spinel, gypsum, hematite, calcite, pyrite, and galena. *(Chip Clark.)*

the abundant minerals of Earth's crust and is the chief constituent of a group of rocks called limestones. Its basic building block, the carbonate ion, $(CO_3)^{2-}$, consists of a carbon atom surrounded by three oxygen atoms in a triangle, as in Figure 2.20a. The carbon shares electrons with oxygens. Groups of carbonate ions are arranged in sheets somewhat like the sheet silicates and are bonded by layers of cations (Figure 2.20b). The sheets of carbonate ions in calcite are separated by layers of calcium ions. The mineral dolomite, $CaMg(CO_3)_2$, another major mineral of crustal rocks, is made up of the same carbonate sheets separated by alternating layers of calcium ions and magnesium ions.

Oxides

Oxide minerals are compounds in which oxygen is bonded to atoms or cations of other elements, usually metallic ions such as iron (Fe^{2+} or Fe^{3+}). Most oxide minerals are ionically bonded, their structures varying with the size of the metallic cations. This group is of great economic importance because it includes the ores of most of the metals, such as chromium and titanium, used in industrial and technological manufacture of metallic materials and devices. Hematite (Fe_2O_3) is a chief ore of iron.

Another of the abundant minerals in this group, spinel, is an oxide of two metals, magnesium and aluminum ($MgAl_2O_4$). Spinel has a closely packed cubic structure and high density (3.6 g/cm^3), reflecting the conditions of high pressure and temperature under which it is formed. Transparent, gem-quality spinel resembles ruby and sapphire and is found in the crown jewels of England and Russia.

Sulfides

The chief ores of many valuable minerals, such as copper, zinc, and nickel, are members of the sulfide group of minerals. This group includes compounds of the sulfide ion, S^{2-}, with metallic cations. The sulfide ion is one in which a sulfur atom has gained two electrons in its outer shell. Most sulfide minerals look like metals, and almost all are opaque. The structures of these minerals are diverse, depending on the way the sulfide anions combine with the metallic cations. The most common sulfide mineral is pyrite (FeS_2), frequently called "fool's gold" because of its yellowish metallic appearance.

Sulfates

In sulfates, sulfur is present as the sulfate ion, a tetrahedron made up of one sulfur atom that has lost six electrons from its outer shell, joined with four oxygen ions (O^{2-}), giving it the formula SO_4^{2-}. The sulfate ion is the basis for a variety of structures. One of the most abundant minerals of this group is gypsum, the primary component of plaster. Gypsum is formed when seawater evaporates. During evaporation, Ca^{2+} and SO_4^{2-}, two ions abundant in seawater, combine and precipitate as layers of sediment, forming calcium sulfate ($CaSO_4 \cdot 2H_2O$). (The dot in this formula signifies that two water molecules are bonded to the calcium and sulfate ions.)

Another calcium sulfate, anhydrite ($CaSO_4$), differs from gypsum in containing no water. Its name is derived from the word *anhydrous,* meaning "free from water." Gypsum is stable at the low temperatures and pressures found at Earth's surface, whereas

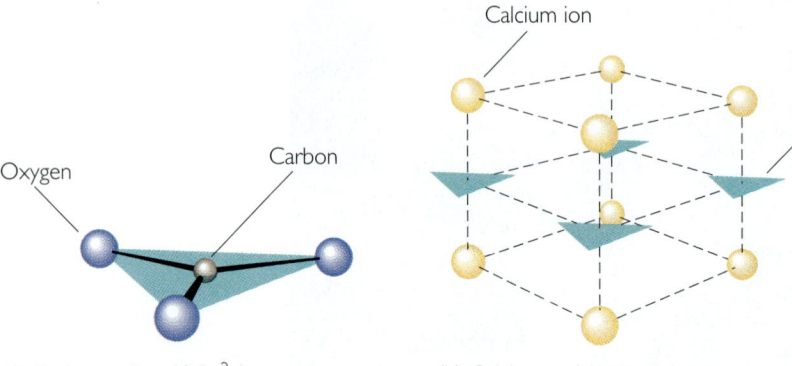

FIGURE 2.20 Carbonate minerals, such as calcite (calcium carbonate, $CaCO_3$), have a layered structure. (a) Top view of the carbonate building block, a carbon atom surrounded in a triangle by three oxygen ions, with a net charge of -2. (b) View of the alternating layers of calcium and carbonate ions.

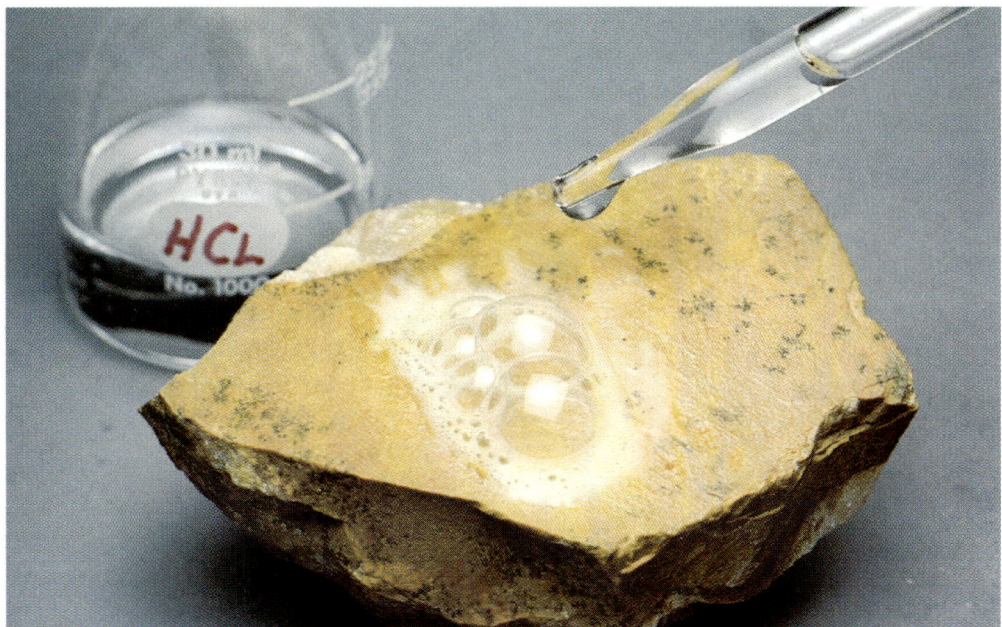

FIGURE 2.21 Acid test. One easy but effective way to identify certain minerals is to drop diluted hydrochloric acid (HCl) on the substance. If it fizzes, indicating the escape of carbon dioxide, the mineral is likely to be calcite. *(Chip Clark.)*

anhydrite is stable at the higher temperatures and pressures of buried sedimentary rocks.

Crystal structure and composition are not useful solely for organizing our knowledge of the minerals of the world. They also explain the physical properties of minerals that make them useful or decorative.

Physical Properties of Minerals

Geologists use the compositions and structures of minerals to understand the origin of the rocks they make up and thus the nature of the geological processes that operate on and in the Earth. This understanding often begins in the field with attempts to identify and classify unknown minerals. At such times, geologists place great reliance on chemical and physical properties that can be observed with relative ease. Most of what we know about the chemical composition of minerals was originally learned through the use of ordinary chemical methods to dissolve minerals and rocks, separate them into their constituent elements, and then measure their weights or volumes. In the nineteenth and early twentieth centuries, geologists used to carry field kits for the rough partial chemical analysis of minerals that would help in identification. One such test is the origin of the phrase "the acid test." It consists of dropping diluted hydrochloric acid (HCl) on a mineral to see if it fizzes. The fizzing indicates that carbon dioxide (CO_2) is escaping, and that means the mineral is likely to be calcite, a carbonate mineral (Figure 2.21).

In the remainder of this chapter we review the physical properties of minerals, many of which contribute to their practical and decorative value. (See Feature 2.1, "What Makes Gems So Special?")

Hardness

Hardness is a measure of the ease with which the surface of a mineral can be scratched. Just as a diamond, the hardest mineral known, scratches glass, so a quartz crystal, which is harder than feldspar, scratches a feldspar crystal. In 1822 Friedrich Mohs, an Austrian mineralogist, devised a scale, now known as the **Mohs scale of hardness,** based on the ability of one mineral to scratch another. At one extreme is the softest mineral (talc); on the other, the

2.1 LIVING ON EARTH

What Makes Gems So Special?

No one can be sure when the first human picked up a crystal of a mineral and kept it for its rare beauty, but we do know that gems were being worn as necklaces and other adornments at the dawn of civilization in Egypt, at least 4000 years ago. These early Egyptians were undoubtedly attracted to the color and play of light on the polished surfaces of such minerals as carnelian, lapis lazuli, and turquoise. Color and luster, or the ability to reflect light, are two of the qualities that still serve to define gemstones. Although the value placed on a gemstone varies from one culture and historical period to another, other required qualities seem to be beauty, transparency, brilliance, durability, and rarity.

Most minerals have some of these remarkable qualities, but the stones considered most precious are ruby, sapphire, emerald, and, of course, diamond. A diamond—geologically speaking, at least—may not be forever, as the advertisers claim, but it is special. Its glitter is unique, as are the play of colors and the sparkle it emits. The source of these qualities is the way diamond refracts, or bends, light. They are enhanced by diamond's remarkable ability to split perfectly along certain directions of the crystal, which diamond cutters use to advantage in carefully cutting gem-quality stones. Diamond's multiple facets (faces superficially similar to crystal faces) can be polished to enhance this sparkle. These facets can be ground only by other diamonds, for this is the hardest mineral known, so hard it can scratch any other mineral and remain undamaged. This mineral's tightly packed crystal structure and strong covalent bonds between carbon atoms give it these characteristics that allow it to be identified with certainty by mineralogists and jewelers.

TABLE 2.3

MOHS SCALE OF HARDNESS

MINERAL	SCALE NUMBER	COMMON OBJECTS
Talc	1	
Gypsum	2	
Calcite	3	Fingernail
Fluorite	4	Copper coin
Apatite	5	
Orthoclase	6	Knife blade
Quartz	7	Window glass
Topaz	8	Steel file
Corundum	9	
Diamond	10	

hardest (diamond) (Table 2.3). The Mohs scale is still one of the best practical means to identify an unknown mineral. With a knife blade and a few of the minerals on the hardness scale, a field geologist can gauge an unknown mineral's position on the scale. If the unknown mineral is scratched by a piece of quartz but not by the knife, for example, it lies between 5 and 7 on the scale.

Recall from our earlier discussion that covalent bonds are generally stronger than ionic bonds. The hardness of any mineral depends on the strength of its chemical bonds; the stronger the bonds, the harder the mineral. Crystal structure varies in the silicate group of minerals, and so does hardness. For example, within the silicates, hardness varies from 1 in talc, a sheet silicate, to 8 in topaz, a silicate with isolated tetrahedra. Most silicates fall in the 5 to 7 range on the Mohs scale. Only sheet silicates are relatively soft, with hardnesses between 1 and 3.

Rubies and sapphires are gem-quality varieties of the common mineral corundum (aluminum oxide), which is widespread and abundant in a number of rock types. Although not as hard as diamond, corundum is extremely hard. Small amounts of impurities produce the intense colors we value. Ruby, for example, is red because of small amounts of chromium, the same substance that gives emeralds their green color.

Less valuable, sometimes called semiprecious, gemstones are topaz, garnet, tourmaline, jade, turquoise, and zircon. Most, like garnet, are common constituents of rocks, occurring mostly as small imperfect crystals with many impurities and poor transparency. But under special conditions, gem-quality garnets form. From time to time, some minerals that are not ordinarily considered gems may enjoy sudden—perhaps temporary—popularity. Hematite (iron oxide) currently enjoys this status, appearing in necklaces and bracelets.

Sapphire (blue) and diamond (colorless) brooch by Fortunato Pio Castellani, Smithsonian Institution, nineteenth century. *(Aldo Tutino/Art Resource.)*

Within groups of minerals having similar crystal structures, increasing hardness is related to factors that also increase bond strength:

- *Size* The smaller the atom or ion, the smaller the distance between the atoms or ions, and thus the stronger the electrical attraction.
- *Charge* The larger the charge of ions, the greater the attraction between ions, and thus the stronger the bond.
- *Packing of atoms or ions* The closer the packing of atoms or ions, the smaller the distance between atoms or ions, and the stronger the bond.

Size is an especially important factor for most metallic oxides and sulfides of metals with high atomic numbers—such as those of gold, silver, copper, and lead. Members of this group are soft, with hardnesses of less than 3, because their metallic cations are so large. Carbonates and sulfates, groups in which the structures are packed less densely, are also soft, with hardnesses of less than 5. In each of these groups, their chemical bonds are reflected in their hardness.

Cleavage

Cleavage is the tendency of a crystal to break along flat planar surfaces. The term is also used to describe the geometric pattern produced by such breakage. Cleavage varies inversely with bond strength—if bond strength is high, cleavage is poor; if bond strength is low, cleavage is good. Because of their strength, covalent bonds generally give poor or no cleavage. Ionic bonds are relatively weak, so they give excellent cleavage.

If the bonds between some of the planes of atoms or ions in a crystal are weak, the mineral can

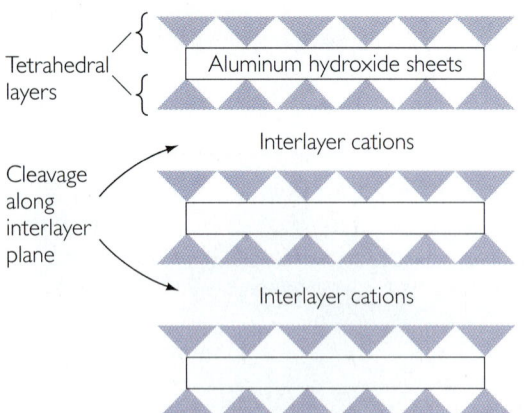

FIGURE 2.22 Cleavage of mica. The diagram shows the cleavage planes in the mineral structure, oriented perpendicular to the plane of the page. Horizontal lines mark the interfaces of silica-oxygen tetrahedral sheets and sheets of aluminum hydroxide bonding the two tetrahedral layers into a "sandwich." Cleavage takes place between composite tetrahedral-aluminum hydroxide sandwiches. The photograph shows thin sheets separating along the cleavage planes. *(Chip Clark.)*

FIGURE 2.23 Example of rhomboidal cleavage in calcite. Calcite can be cleaved by a light hammer blow on a chisel oriented parallel to one of its planes. *(Chip Clark.)*

be made to split along those planes. Muscovite, a mica and a sheet silicate, breaks along smooth, lustrous, flat, parallel surfaces, forming thin transparent sheets less than a millimeter thick. Mica's excellent cleavage is the result of weakness of the bonds between the sandwiched layers of cations and tetrahedral silica sheets (Figure 2.22).

Cleavage is classified according to two primary sets of characteristics: the number of planes and pattern of cleavage, and the quality of surfaces and ease of cleaving.

NUMBER OF PLANES; PATTERN OF CLEAVAGE
The number of planes and patterns of cleavage are identifying hallmarks of many rock-forming minerals. Muscovite has only one plane of cleavage, but calcite and dolomite have three excellent cleavage directions that give them a rhomboidal shape (Figure 2.23).

A crystal's structure determines its cleavage planes and its crystal faces. Crystals have fewer cleavage planes than possible crystal faces. Faces may be formed along any of numerous planes defined by rows of atoms or ions. Cleavage occurs along any of those planes across which the bonding is weak. All crystals of a mineral exhibit its characteristic cleavage, whereas only some crystals display particular faces.

Galena (lead sulfide, PbS) and halite (sodium chloride, NaCl) cleave along three planes, forming perfect cubes. Distinctive angles of cleavage help identify two important groups of silicates, the pyrox-

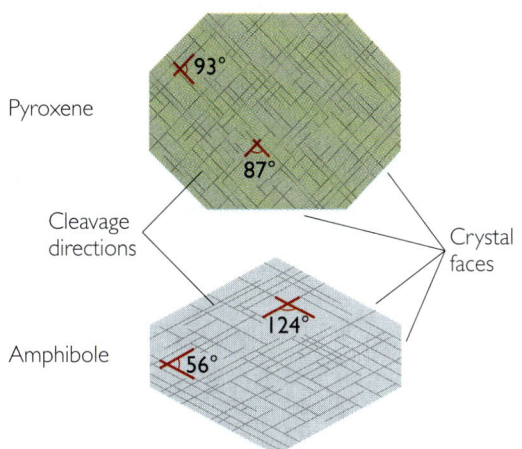

FIGURE 2.24 Comparison of cleavage directions and typical crystal faces in pyroxene and amphibole. These two minerals often look very much alike; but their angles of cleavage differ. These angles are frequently used to identify and classify them.

cleavage. This absence of a tendency to cleave is found in most framework silicates and in silicates with isolated tetrahedra.

Fracture

Fracture is the tendency of a crystal to break along irregular surfaces other than cleavage planes. All minerals show fracture, either across cleavage planes or in any direction in minerals, such as quartz, with no cleavage. Fracture is related to the way bond strengths are distributed in directions that cut across crystal planes. Breakage of these bonds results in irregular fractures. Fractures may be *conchoidal*, showing smooth, curved surfaces like those of a thick piece of broken glass. A common fracture surface with an appearance like split wood is described as *fibrous* or *splintery*. The shape and appearance of many kinds of irregular fractures depend on the particular structure and composition of the mineral.

enes and amphiboles, that otherwise often look alike (Figure 2.24). Pyroxenes have a single-chain linkage, and they are bonded so that their cleavages are almost at right angles (93°) to each other. In cross section, the cleavage pattern of pyroxene is nearly a square. In contrast, amphiboles, the double chains, bond to give two cleavage directions, at 56° and 124° to each other. They produce a diamond-shaped cross section.

QUALITY OF SURFACES; EASE OF CLEAVING A mineral's cleavage is assessed as perfect, good, or fair, according to the quality of surfaces produced and the ease of cleaving. Muscovite can be cleaved easily, producing extremely high-quality, smooth surfaces; its cleavage is *perfect*. The single- and double-chain minerals (pyroxenes and amphiboles, respectively) show *good* cleavage. Although these minerals break easily along the cleavage plane, they also break across it, producing cleavage surfaces that are not as smooth as those of mica. *Fair* cleavage is shown by the ring silicate beryl. Beryl's cleavage is more irregular, and the mineral breaks relatively easily along directions other than cleavage planes.

Many minerals are so strongly bonded that they lack even fair cleavage. Quartz, one of the commonest, is a framework silicate; it is so strongly bonded in all directions that it breaks only along irregular surfaces. Garnet, an isolated tetrahedral silicate, is also bonded strongly in all directions and so has no

Luster

The way the surface of a mineral reflects light gives it a characteristic **luster**. Mineral lusters are described by the terms listed in Table 2.4. Luster quality is controlled by the kinds of atoms present and their bonding, both of which affect the way light passes

TABLE 2.4

MINERAL LUSTER

Metallic	Strong reflections produced by opaque substances
Vitreous	Bright, as in glass
Resinous	Characteristic of resins, such as amber
Greasy	The appearance of being coated with an oily substance
Pearly	The whitish iridescence of such materials as pearl
Silky	The sheen of fibrous materials such as silk
Adamantine	The brilliant luster of diamond and similar minerals

through or is reflected by the mineral. Ionically bonded crystals tend to be glassy, or vitreous, but covalently bonded materials are more variable. Many tend to have an adamantine luster, like that of diamond. Metallic luster is shown by pure metals, such as gold, and sulfides, such as galena (lead sulfide, PbS). Pearly luster is the result of multiple reflections of light from planes beneath the surfaces of translucent minerals, such as the mother-of-pearl inner surfaces of many clam shells, which are made of the mineral aragonite. Luster quality, although an important criterion for field classification, depends heavily on visual perception of reflected light. Textbook descriptions fall short of the actual experience of holding the mineral in your hand.

Color

The **color** of a mineral is imparted by light—either transmitted through or reflected by crystals, irregular masses, or a streak. **Streak** is the name given to the color of the fine deposit of mineral dust left on an abrasive surface, such as a tile of unglazed porcelain, when a mineral is scraped across it. Such tiles are called *streak plates* (see Figure 2.25). A streak plate is a good diagnostic tool because the uniform small grains of mineral that occur in the powder on the ceramic tile permit a better analysis of color than does a mass of the mineral. A mass formed of hematite (Fe_2O_3), for example, may be black, red, or brown, but this mineral will always leave a trail of reddish-brown dust on a streak plate.

Color is determined both by the kinds of ions found in the pure mineral and by trace impurities. Color is a complex and as yet not fully understood property of minerals; its details, other than those given below, are beyond the scope of this book.

IONS AND COLOR OF MINERALS The color of pure substances depends on the presence of certain ions, such as iron or chromium, which strongly absorb portions of the light spectrum. Olivine containing iron, for example, absorbs all colors except green, which it reflects, so we see this type of olivine as green. We see pure magnesium olivine as white (transparent and colorless). Most ionically bonded pure minerals whose ions have full, stable outer electron shells, such as halite, are colorless.

TRACE IMPURITIES AND MINERAL COLOR All natural minerals contain impurities. In the past few decades, new instruments have made it possible to measure even very small quantities of some elements—as little as a billionth of a gram in some cases. Elements that make up much less than 0.1 percent of a mineral are reported as "traces," and many of these are called trace elements.

Some trace elements are useful in interpreting the origins of the minerals in which they are found. Others, such as the trace amounts of uranium in some granites, contribute to local natural radioactivity. Still others, such as small dispersed flakes of hematite that color a feldspar crystal brownish or reddish, are notable because they give a general color to an otherwise colorless mineral. Many of the gem varieties of minerals, such as emerald (green beryl) and sapphire (blue corundum), get their color from trace impurities dissolved in the solid crystal (see Feature 2.1). Emerald derives its color from chromium; the source of sapphire's blue is iron and titanium.

The color of a mineral may be distinctive, but it is not the most reliable clue to its identity. Some minerals always show the same color; others may have a range of colors. Many minerals show a characteristic color only on freshly broken surfaces or only on weathered surfaces. Some—precious opals, for example—show a stunning display of colors on reflecting surfaces. Others change color slightly with a change in the angle of the light shining on their surface.

FIGURE 2.25 Hematite may be black, red, or brown, but it always leaves a reddish brown streak when scratched along a ceramic plate. *(Breck P. Kent.)*

TABLE 2.5
PHYSICAL PROPERTIES OF MINERALS

PROPERTY	RELATION TO COMPOSITION AND CRYSTAL STRUCTURE
Hardness	Strong chemical bonds give high hardness. Covalently bonded minerals are generally harder than ionically bonded minerals.
Cleavage	Cleavage is poor if bond strength in crystal structure is high and is good if bond strength is low. Covalent bonds generally give poor or no cleavage; ionic bonds are weak, so give excellent cleavage.
Fracture	Type is related to distribution of bond strengths across irregular surfaces other than cleavage planes.
Luster	Tends to be glassy for ionically bonded crystals, more variable for covalently bonded crystals.
Color	Determined by kinds of atoms and trace impurities. Many ionically bonded crystals are colorless. Iron tends to color strongly.
Streak	Color of fine powder is more characteristic than that of massive mineral because of uniform small size of grains.
Density	Depends on atomic weight of atoms and their closeness of packing in crystal structure. Iron minerals and metals have high density; covalently bonded minerals have more open packing and so have lower density.

Specific Gravity and Density

The difference in weight between a piece of hematite iron ore and a piece of sulfur of the same size is easily felt when the two pieces are hefted. A great many common rock-forming minerals, however, are too similar in density for such simple tests as hefting. **Density** is mass per unit volume (usually expressed in grams per cubic centimeter—g/cm^3). Scientists therefore needed some method that would make it easy to measure this property of minerals. A standard measure of density is **specific gravity,** which is the weight of a mineral in air divided by the weight of an equal volume of pure water at 4°C.

Density depends on the atomic weight of the mineral's ions and the closeness with which they are packed in a mineral's crystal structure. Consider the iron oxide magnetite, with a density of 5.2 g/cm^3. This high density results partly from the high atomic weight of iron and partly from the closely packed structure magnetite shares with the other members of the spinel group of minerals (see page 44). The density of the iron silicate olivine, 4.4 g/cm^3, is lower than that of magnetite for two reasons. First, the atomic weight of silicon, one of the elements from which olivine is formed, is lower than that of iron. Second, this olivine has a more openly packed structure than that of the spinel group. The density of the magnesium olivine is even lower, 3.32 g/cm^3, because magnesium's atomic weight is much lower than that of iron. Table 2.5 summarizes the physical properties of minerals.

Increases of density caused by pressure affect the way minerals transmit light, heat, and earthquake waves. Experiments at extremely high pressures have shown that olivine converts to the denser structure of the spinel group at pressures corresponding to a depth of 400 km. At a greater depth, 670 km, mantle materials are further transformed to silicate minerals with the even more densely packed structure of the mineral perovskite (calcium titanate, $CaTiO_3$). Because of the huge volume of the lower mantle, silicate with the perovskite structure is probably the most abundant mineral in the Earth as a whole (see page 39). Some perovskite minerals have been synthesized to be high-temperature semiconductors,

2.2 LIVING ON EARTH

Asbestos: Health Hazard, Overreaction, or Both?

Mention of asbestos, once used extensively as a fireproof insulator and flame retardant in plaster, ceiling and floor tile, and automobile insulation, has come to provoke fear in the last two decades. During that time, asbestos has been linked to several fatal lung diseases—such as asbestosis (characterized by progressive lung stiffening and difficulty in breathing) and lung cancer, including a specific cancer called mesothelioma, which attacks the lining of the lung. The specific link seems to be heavy exposure to certain minerals lumped under the commercial name *asbestos* or exposure to them over a prolonged period of time. The exact way these substances cause the diseases is still not well understood, but the sharp fibrous crystal habit of some of the minerals has been implicated.

The health problems associated with exposure to asbestos came to public attention when lawyers representing workers and their families brought class-action lawsuits against some of the major companies that fabricated asbestos products. The lawsuits accused the companies of responsibility for the deaths and disabilities of large numbers of people formerly employed in asbestos factories. In the wake of settlements awarded as a result of these lawsuits, some of the companies went bankrupt. They left behind them a public intensely concerned over asbestos-containing materials in place in schools, hospitals, and other public and private buildings. Many states now require disclosure of such materials during negotiations for the sale of private homes, for example.

Although research specialists who identify and treat lung diseases have been vocal in their warnings to the public, other responsible scientists think the frenzied concern about all forms of asbestos is an overreaction. At the center of this debate lies mineralogy.

Six distinct minerals are lumped under the commercial term *asbestos:* chrysotile, a sheet silicate member of the serpentine group;

Asbestos (chrysotile). Fibers are readily combed from the solid mineral. *(Runk/Schoenberger/Grant Heilman Photography.)*

crocidolite, a double-chain silicate member of the amphibole group; and four other double-chain silicates in the serpentine group. Although crocidolite does form sharp fibers, many of the other minerals do not. Crocidolite has been heavily implicated in lung diseases. U.S. government regulations nevertheless apply to all of these minerals.

Physicians and mineralogists vary widely in their opinions on whether all forms of asbestos should be eliminated from buildings and factories. Any decision must consider several facts. First, heavy exposure to crocidolite is dangerous, especially for smokers. Smokers may develop lung diseases without any exposure to asbestos, but smokers are much more prone to asbestos-related lung diseases than are non-smokers. Prolonged occupational exposure to some other forms of asbestos, as by workers in asbestos factories, may also create a danger.

Abundant evidence exists, however, to indicate that people exposed to moderate amounts of chrysotile, the most commonly used asbestos in North America, for long periods of time show no asbestos-related lung disease. The lack of a correlation between exposure to specific asbestos minerals, verified by laboratory analysis, and the occurrence of specific lung diseases also contributes to confusion about the danger of exposure to asbestos. The situation is particularly ambiguous in respect to the danger posed to the general population by asbestos in large public buildings.

Many medical scientists and mineralogists doubt that we need to spend the $50 billion to $150 billion it would cost to clean up relatively harmless chrysotile and the four forms of amphibole that do not form sharp fibers. Before any decision can be made with certainty, we need more mineralogical evaluation and additional medical studies to determine the nature of the problem confronting us.

which conduct electricity without loss and may have large commercial potential. Mineralogists experienced with the natural material helped unravel the structure of these newly created materials. Temperature also affects density: the higher the temperature, the more open and expanded the structure and thus the lower the density.

Crystal Habit

A mineral's **crystal habit** is the shape in which its individual crystals or aggregates of crystals grow. Crystal habits are often named after common geometric shapes, such as blades, plates, and needles. Some minerals have such a distinct and characteristic crystal habit that they are easily recognizable, such as quartz's six-sided column topped by a pyramidlike set of faces. These shapes reflect not only the planes of atoms or ions in the mineral's crystal structure but also the typical speed and direction of crystal growth. Thus a needlelike crystal is one that grows very quickly in one direction and very slowly in all other directions. In contrast, a plate-shaped crystal (often referred to as *platy*) grows fast in all directions that lie perpendicular to its single direction of slow growth. Fibrous crystals take shape as multiple long, narrow fibers, essentially aggregates of long needles. *Asbestos* is a generic name for a group of silicates with a more or less fibrous habit that allows the crystals to become embedded after having been inhaled into the lungs (see Feature 2.2, "Asbestos: Health Hazard, Overreaction, or Both?").

In summary, minerals exhibit a variety of physical and chemical properties that result from their chemical compositions and atomic structures. Many of these properties are useful to the mineralogist or geologist for purposes of identification or classification. Geologists study the compositions and structures of minerals to understand the origin of the rocks in which they are found, and thus the nature of the geological processes that operate on and in the Earth.

In the chapters that follow, we will refer to various minerals in one geological context or another. Appendix 3, just before the glossary at the end of this book, lists the properties of the most common minerals in Earth's crust. We urge you to consult this appendix when you want more information about these minerals.

Summary

What is a mineral? Minerals, the building blocks of rocks, are naturally occurring inorganic solids with specific crystal structures and chemical compositions that are fixed or vary within a defined range. A mineral is constructed of atoms, the small units of matter that combine in chemical reactions. An atom is composed of a nucleus of protons and neutrons, surrounded by electrons traveling in orbitals around the nucleus. The atomic number of an element is the number of protons in its nucleus, and its atomic weight is the sum of the masses of its protons and neutrons.

How are atoms combined to form the crystal structures of minerals? Chemical substances react with each other to form compounds either by gaining or losing electrons to become ions or by sharing electrons. Either way, the atoms achieve stable configurations of electron shells. The atoms or ions of a substance are held together by ionic and covalent bonds formed by electrostatic attraction between nuclei and electrons of the constituent elements. When a mineral crystallizes, atoms or ions come together in the proper proportions to form a crystal structure, which is an orderly, three-dimensional geometric array in which the basic arrangement is repeated in all directions.

What are the major rock-forming minerals? Silicates, the most abundant minerals in Earth's crust, are crystal structures built of silicate tetrahedra linked in various ways. Tetrahedra may be isolated (olivines) or in rings (cordierite), single chains (pyroxenes), double chains (amphiboles), sheets (micas), or frameworks (feldspars). Carbonate minerals are made of carbonate ions bonded to calcium and/or magnesium. Oxide minerals are compounds of oxygen and metallic elements. Sulfide and sulfate minerals are structures made up of sulfur atoms in combination with metallic elements.

What are the physical properties of minerals? Physical properties, which reflect the compositions and structures of minerals, include hardness, or the ease with which a mineral surface is scratched; cleavage, or the ability of a mineral to split or break along flat surfaces; fracture, or the way in which minerals break along irregular surfaces; luster, or the nature of a mineral's reflection of light; color, imparted by either transmitted or reflected light to crystals, irregular masses, or a streak (the color of a fine powder); density, or the mass per unit volume; and crystal habit, or the shapes of individual crystals or aggregates.

Key Terms and Concepts

mineralogy (p. 28)
mineral (p. 28)
atom (p. 29)
nucleus (p. 29)
proton (p. 29)
neutron (p. 29)
electron (p. 30)
shell (p. 30)
atomic number (p. 30)
atomic mass (p. 30)
isotope (p. 30)

chemical reaction (p. 30)
ion (p. 31)
cation (p. 31)
anion (p. 31)
electron sharing (p. 32)
ionic bond (p. 35)
covalent bond (p. 35)
metallic bond (p. 35)
crystallization (p. 36)
crystal (p. 37)
precipitate (p. 38)

polymorph (p. 39)
hardness (p. 45)
Mohs scale of hardness (p. 45)
cleavage (p. 47)
fracture (p. 49)
luster (p. 49)
color (p. 50)
streak (p. 50)
density (p. 51)
specific gravity (p. 51)
crystal habit (p. 53)

Mineral Names to Remember

(See Appendix 3 for specific mineral properties.)

amphibole	garnet	muscovite
anhydrite	gold	olivine
aragonite	graphite	opal
calcite	gypsum	perovskite
clay mineral	halite	pyrite
corundum	hematite	pyroxene
diamond	hornblende	quartz
dolomite	kaolinite	spinel
enstatite	lead	talc
feldspars	magnetite	
galena	mica	

Exercises

1. Define a mineral.
2. What is the difference between an atom and an ion?
3. Draw the atomic structure of sodium chloride.
4. What are two types of chemical bonds?
5. What are the two polymorphs of carbon?
6. List the basic structure of silicate minerals.
7. How does the cleavage of mica reflect its atomic structure?
8. Name three groups of minerals, other than silicates, based on their chemical composition.
9. What two factors account for the densities of mantle minerals?

Thought Questions

(Consult Appendix 3 for specific mineral properties.)

1. Joan and Alex are comparing rubies. Joan's is natural and Alex's is synthetic. Are both of them looking at minerals? Why, or why not?
2. Hydrogen (H), the lightest element, has an atomic number of 1 and an atomic weight of 1.008 in nature. What does this information tell you about possible isotopes of hydrogen?
3. Draw a simple diagram to show how silicon and oxygen in silicate minerals share electrons. Model your diagram on Figure 2.5.
4. An isotope of the element aluminum has 13 protons and an atomic weight of 27. Use the periodic table of elements to find its atomic number and then draw its electron shells and its nucleus, using Figure 2.3 as a model.
5. Use iron and magnesium in silicate minerals to illustrate cation substitution.
6. Diopside, a pyroxene, has the formula $(Ca,Mg)_2Si_2O_6$. What does this tell you about its crystal structure and cation substitution?
7. Oxygen exists as three isotopes with atomic weights of exactly 16, 17, and 18. The atomic weight of oxygen found in nature is approximately 16. What does this information tell you about the relative abundance of the three isotopes in nature?
8. In some places in bodies of granite, we can find very large crystals, some as much as a meter across, yet these crystals tend to have few crystal faces. What can you deduce about the conditions under which these large crystals grew?

9. What physical properties of sheet silicates are related to their crystal structure and bond strength?
10. How might you identify and differentiate between a single- and a double-chain silicate?
11. What physical properties would make calcite a poor choice for a good gemstone?
12. Choose two minerals from Appendix 3 that you think might make good abrasive or grinding stones for sharpening steel, and describe the physical property that causes you to believe they would be suitable for this purpose.
13. Aragonite, with a density of 2.9 g/cm^3, has exactly the same chemical composition as calcite, with a density of 2.7 g/cm^3. Other things being equal, which of these two minerals is more likely to have formed under high pressure?
14. What properties of talc make it suitable for face and body powder?

Short-Term Team Project: Asbestos

Children were in danger. News reports of asbestos in the New York City schools touched off a wave of public outrage, and the Board of Education postponed the start of the 1993 school year for three weeks to complete an asbestos-removal program.

Reports of asbestos in public buildings almost always focus on the risk of lung disease. Seldom, however, do these reports include interviews with mineralogists. As a result, the public knows too little about asbestos to ask the right questions and make an informed judgment.

You now have the chance to educate the public. Working in a team of four students over the next two weeks, prepare a half-hour radio program representing a variety of perspectives on the asbestos issue. The program will show how useful a basic knowledge of mineralogy can be in deciding important policy questions. Assemble a cast of experts with opposing positions on the issue. From what disciplines would they come? With what arguments and facts would they support their positions? Write a script for the program and deliver an oral summary of your list of experts and their positions.

Suggested Readings

Berry, L. G., B. Mason, and R. V. Dietrich. 1983. *Mineralogy*, 2nd ed. San Francisco: W. H. Freeman.

Dietrich, R. V., and B. J. Skinner. 1990. *Gems, Granites, and Gravels.* Cambridge: Cambridge University Press.

Keller, P. C. 1990. *Gemstones and Their Origins.* New York: Chapman & Hall.

Klein, C., and C. S. Hurlbut, Jr. 1993. *Manual of Mineralogy*, 21st ed. New York: Wiley.

McQuarrie, D. A., and P. A. Rock. 1991. *General Chemistry*, 3rd ed. New York: W. H. Freeman.

Prinz, M., G. Harlow, and J. Peters. 1978. *Simon & Schuster's Guide to Rocks and Minerals.* New York: Simon & Schuster.

Internet Sources

Mineral Gallery
  http://mineral.galleries.com/
This commercial site provides a data base to search for minerals by name, chemical composition, class (oxides, silicates, etc.), and groupings (birthstones, gemstones, etc.). A search produces an image of the mineral, data on composition, characteristics, properties, and uses. The physical properties of minerals are also explained.

Mineralogy
  http://un2sg1.unige.ch/www/athena/mineral/mineral.html
This site, based in Switzerland, provides images and a data base that can be searched for minerals by name or chemical composition.

Smithsonian Gem & Mineral Collection
🛈 **http://galaxy.einet.net/images/gems/gems-icons.html**
This site features images and descriptions of outstanding mineral specimens in the Smithsonian's collection, including the Hope Diamond and the Star of Bombay sapphire.

Minerals in Thin Sections
🛈 **http://www.science.ubc.ca/~geol202/s/cgi-bin/mineral.cgi**
Maintained by the University of British Columbia, this site includes photomicrographs of minerals in plane and polarized light and lists their optical properties.

USGS Minerals Information
🛈 **http://minerals.er.usgs.gov/minerals/**
The U.S. Geological Survey provides information under such headings as What's New, Publications and Information Products (Mineral Year Book, etc.), Gemstone Production, World Gold, and more.

3

Sandstone formations, painted cliffs, Tasmania, Australia. These sandstones represent a former cycle of uplift, erosion, and sedimentation. Erosion of the sandstone cliffs is part of the current cycle. (John Cancalosi/DRK.)

Rocks: Records of Geologic Processes

What determines the appearance of the rocks we encounter? They vary in color, in the sizes of their crystals or grains, and in the kinds of minerals that make them up. Along a road cut, for example, we may find a black, homogeneous rock, its constituent particles—volcanic glass and crystals of pyroxene and feldspar—too small to be seen with the naked eye. Near it may be a brownish rock; this one, transformed by heat and pressure deep in the Earth, has abundant large glittering crystals of mica and some grains of quartz and feldspar. Overlying both the black rock and the brown one may be the remains of a former beach, horizontal layers of light-brown rock that appear to be made up of sand grains cemented together.

The appearance of these rocks is determined partly by their mineralogy and partly by their texture. Mineralogy, or the relative proportions of a rock's constituent minerals, helps determine its appearance and other properties, as you will recall from Chapter 2. So does the rock's **texture,** or the sizes and shapes of its mineral grains and crystals and the way they are put together. These grains or crystals, only a few millimeters in diameter in most rocks, are categorized as *coarse* (if they are large enough to be seen with the naked eye) or *fine* (if they are not). Mineral grains or crystals also vary in shape: they may be needle-shaped, flat, platy, or equant (about the same dimension in all directions, like a sphere or a cube). These variations in mineralogy and texture combine to produce larger features that define individual rocks.

The mineralogy and texture that determine a rock's appearance are themselves determined by the rock's geologic origin—where and how it was made (Figure 3.1). The dark rock in our road cut, called *basalt,* was formed by a volcanic eruption; its mineralogy and texture depend on the chemical composition of rocks that were melted deep in Earth's interior and on the nature of the eruption—whether it was explosive or a quieter lava flow. All rocks that were formed by the solidification of molten rock are called **igneous rocks.**

The light-brown layered rock of the road cut, a *sandstone,* was formed as sand particles accumulated, perhaps on a beach, and eventually were covered over, buried, and cemented together to form a rock. All rocks that were formed as the burial products of layers of sediments—such as sand, mud, and calcium carbonate shells—whether they were laid down on the land or under the sea, are called **sedimentary rocks.**

The brownish rock of our road cut, a *schist,* contains crystals of mica, quartz, and feldspar and was formed deep in Earth's crust as high temperatures and pressures transformed the mineralogy and texture of a buried sedimentary rock. All rocks that are formed by transformations of preexisting rocks in the solid state under the influence of high pressure and temperature are called **metamorphic rocks.**

Understanding rock properties and reasoning from them to deduce their geologic origins is the primary aim of a geologist. Such deductions, as you learned earlier, not only are essential to the ongoing process of understanding the planet we live on but also are important sources of information about fuel reserves and solutions to environmental problems.

FIGURE 3.1 The minerals and textures of the three great rock groups are formed in different places in the Earth by different geologic processes. As a result, geologists use mineralogical and chemical analyses of rocks to determine the origins of rocks and the processes that formed them. The igneous rock shown here is basalt, formed by volcanism, in Yellowstone National Park. *(Willard Clay.)* The sedimentary rock is sandstone, formed from sand particles, in Sedona, Arizona. *(Michael Long/Visuals Unlimited.)* The metamorphic rock is schist, produced by very high temperatures and pressures acting on accumulated crystals of mica and other minerals, on the Blue Ridge Parkway in North Carolina. *(Gary Meszaros.)*

Knowing that oil, for example, is formed in certain kinds of sedimentary rocks that are rich in organic remains of biological origin, we can explore for new oil reserves more intelligently. Similarly, our knowledge of the properties of rocks is necessary for the discovery of other useful and economically valuable mineral and energy resources, such as gas, coal, and metal ores.

Understanding how rocks form also guides us to the solution of environmental problems. Will this rock be prone to earthquake-triggered landslides? How may it transmit polluted waters in the ground? The underground storage of radioactive and other wastes depends on the analysis of the rock to be used as a repository.

If rocks are clues to many of the things we want to know about the Earth, how do we go about interpreting them? We need a key, just as historians needed the Rosetta stone to crack the "code" of Egyptian hieroglyphics before they could read the inscriptions on temples and tombs. The first step in finding this key is to recognize the various kinds of rocks. The second step is to understand what their characteristics tell us about the surface and subsurface conditions under which they formed.

This chapter gives an overview of how geologists interpret the three great families of rock—igneous, sedimentary, and metamorphic. We see what the appearance, texture, mineralogy, and chemical composition of a rock reveal about how and where it formed. We look at how rock patterns found in subsurface drilling and in outcrops can help us reconstruct geologic history. Finally, we trace the rock cycle—the set of processes that convert each type of rock into the other two—and see how these processes are all driven by plate tectonics.

Igneous Rocks

Igneous rocks (from Latin *ignis*, fire) form by crystallization from a magma, a mass of melted rock that originates deep in the crust or upper mantle, where temperatures reach the 700°C or more needed to melt most rocks. When magmas cool slowly in the interior, microscopic crystals start to form. As the magma cools below the melting point, some of these crystals have time to grow to several millimeters or larger before the whole mass is crystallized as a coarse-grained igneous rock. But when a magma erupts from a volcano onto Earth's surface, it cools and solidifies so rapidly that individual crystals have no time for gradual growth. In that case, many tiny crystals form simultaneously, and the result is a fine-grained igneous rock. Geologists distinguish two major types of igneous rocks—intrusive and extrusive—on the basis of the sizes of their crystals.

Intrusive Igneous Rocks

Intrusive igneous rocks are formed by slowly crystallizing magmas that have intruded rock masses deep in the interior of Earth. They can be recognized by their interlocking large crystals, which grew slowly as the magma gradually cooled (Figure 3.2). Magmas cool slowly in Earth's interior because they

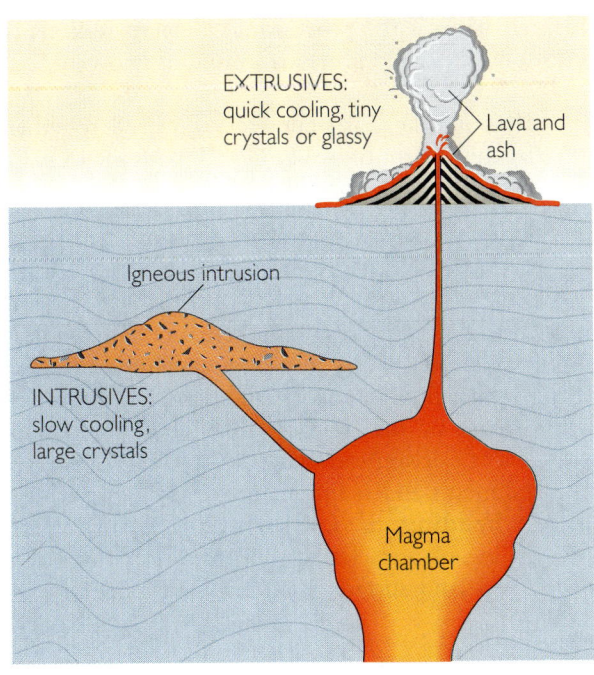

Basalt: an igneous extrusive

Granite: an igneous intrusive

FIGURE 3.2 *Extrusive igneous rocks* are formed when magma erupts at the surface, rapidly cooling to fine ash or lava and forming tiny crystals. The resulting rock, such as the basalt sample here, is finely grained or has a glassy texture. *Intrusive igneous rocks* crystallize when molten rock intrudes into unmelted rock masses deep in Earth's crust. Large crystals grow during the slow cooling process, producing coarsely grained rocks such as the granite sample shown here.

invade rock masses that conduct heat slowly; in addition, the temperatures of some of the rock masses may not be a great deal cooler than the magma itself. *Granite* is an intrusive igneous rock.

Extrusive Igneous Rocks

Rocks such as *basalt,* which form from rapidly cooled magmas that erupt at the surface, are called **extrusive igneous rocks.** They are easily recognized by their glassy or fine-grained texture (see Figure 3.2). Extrusive igneous rocks are formed by **volcanism**—the processes that form volcanoes. These rocks may consist of almost instantaneously crystallized ash particles that were blown high into the atmosphere as a volcano erupted, or of lavas, which flow as liquids for some distance on Earth's surface before they solidify.

Common Minerals

Most of the minerals of igneous rocks are silicates, partly because silicon is so abundant in the Earth and partly because many silicate minerals, but few oxide minerals, melt at the temperatures and pressures reached in lower parts of the crust and the mantle. As Table 3.1 indicates, the common silicate minerals found in igneous rocks include quartz, feldspar, mica, pyroxene, amphibole, and olivine, the minerals that typify the various crystal structures we described in Table 2.2.

TABLE 3.1
SOME COMMON MINERALS OF IGNEOUS, SEDIMENTARY, AND METAMORPHIC ROCKS

IGNEOUS ROCKS	SEDIMENTARY ROCKS	METAMORPHIC ROCKS
Quartz★	Quartz★	Quartz★
Feldspar★	Clay minerals★	Feldspar★
Mica★	Feldspar★	Mica★
Pyroxene★	Calcite	Garnet★
Amphibole★	Dolomite	Pyroxene★
Olivine★	Gypsum	Staurolite★
	Halite	Kyanite★

Asterisks indicate that a mineral is a silicate.

SEDIMENTARY ROCKS

Sediments, the precursors of sedimentary rocks, are found on Earth's surface as layers of loose particles, such as sand, silt, and shells of organisms. Particles such as sand grains and pebbles form at the surface of the Earth as rocks undergo **weathering**—that is, they are broken up into fragments of various sizes. The fragmented rock particles created by weathering are then transported by **erosion**—the set of processes that loosen soil and rock and move them downhill or downstream, where they are laid down as layers of sediment (see Figure 3.3). Weathering and erosion produce two types of sediments:

- **Clastic sediments** are physically deposited sedimentary particles, such as grains of quartz and feldspar derived from a weathered granite. (*Clastic* is derived from the Greek word *klastos*, meaning "broken.") These sediments are laid down by running water, wind, and ice, in the process forming layers of sand, silt, and gravel.

- **Chemical and biochemical sediments** are new chemical substances that form by precipitation when some of a rock's components dissolve during weathering and are carried in river waters to the sea. These sediments include layers of minerals such as halite (sodium chloride) and calcite (calcium carbonate, most frequently found in the form of shells).

From Sediment to Solid Rock

Lithification is the process that converts sediments into solid rock, and it occurs in one of two ways:

- By *compaction,* as grains are squeezed together by the weight of overlying sediment into a mass denser than the original.

- By *cementation,* as minerals precipitate around deposited particles and bind them together.

Sediments are compacted and cemented after burial under additional layers of sediment. Thus sandstone forms by the lithification of sand particles, and limestone forms by the lithification of shells and other particles of calcium carbonate.

Sediments and sedimentary rocks are characterized by **bedding,** the formation of parallel layers by the settling of particles to the bottom of the sea, a river, or a land surface. Bedding may reflect varia-

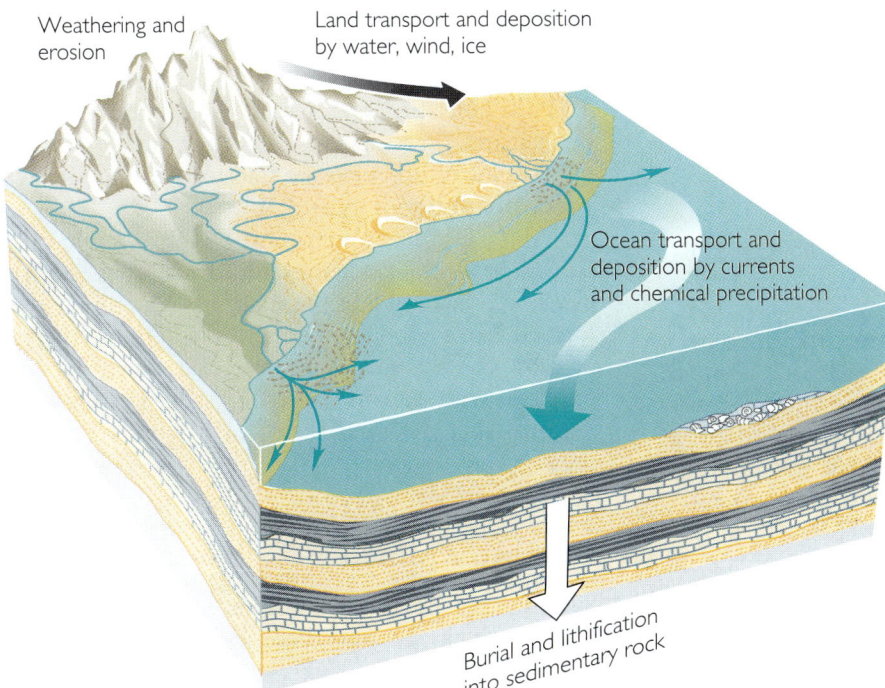

FIGURE 3.3 Weathering breaks down rock into smaller particles that are then carried downhill and downstream by erosion to be deposited as layers of sediment. Other sediment is produced by chemical precipitation. As layers accumulate and are buried deeper and deeper, they lithify, hardening into sedimentary rock.

tions in mineralogy, as when sandstone is interbedded with limestone, or differences in texture, as when a coarse-grained sandstone becomes interbedded with a fine-grained one.

Formed by surface processes, sedimentary rocks cover much of Earth's land surface and seafloor. Although most rocks found at the Earth's surface are sedimentary, they form only a thin layer atop the igneous and metamorphic rocks that make up the main volume of the crust (Figure 3.4).

Common Minerals

The common minerals of clastic sediments are silicates, reflecting the dominance of silicate minerals in rocks that weather to form sedimentary particles (see Table 3.1). Thus the most abundant minerals in clastic sedimentary rocks are quartz, feldspar, and clay minerals.

The most abundant minerals of chemically or biochemically precipitated sediments are carbonates.

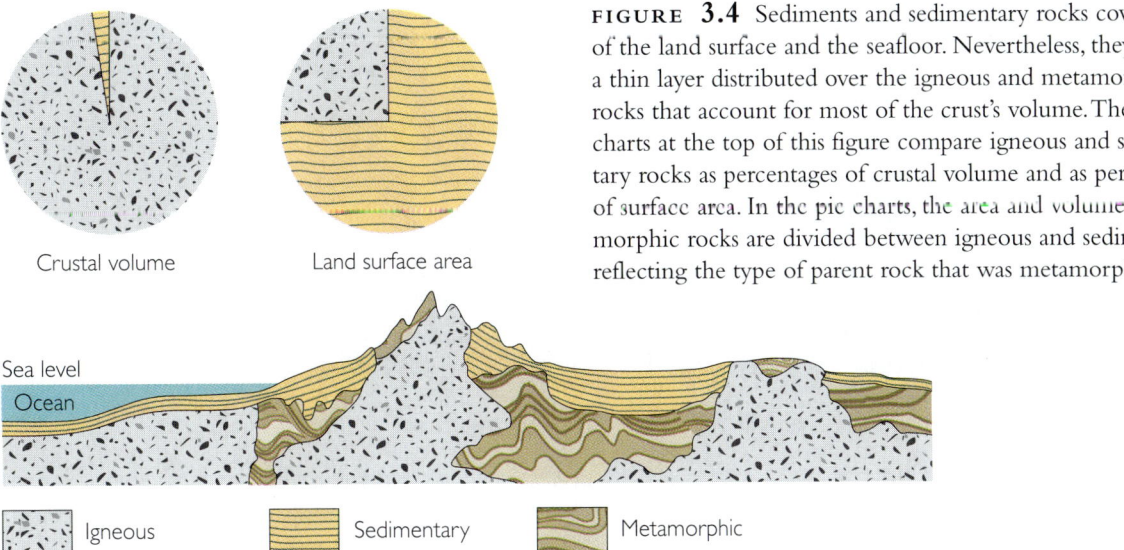

FIGURE 3.4 Sediments and sedimentary rocks cover much of the land surface and the seafloor. Nevertheless, they are only a thin layer distributed over the igneous and metamorphic rocks that account for most of the crust's volume. The pie charts at the top of this figure compare igneous and sedimentary rocks as percentages of crustal volume and as percentages of surface area. In the pie charts, the area and volume of metamorphic rocks are divided between igneous and sedimentary, reflecting the type of parent rock that was metamorphosed.

One is calcite, the main constituent of limestone. Another is dolomite, which is also found in limestone and is a calcium-magnesium carbonate formed by precipitation during lithification. Two others—gypsum and halite—form by chemical precipitation as seawater evaporates.

Metamorphic Rocks

Metamorphic rocks take their name from the Greek words for "change" (*meta*) and "form" (*morphe*). These rocks are produced when high temperatures and pressures deep in the Earth cause any kind of rock—igneous, sedimentary, or other metamorphic rock—to change its mineralogy, texture, or chemical composition while maintaining its solid form. The temperatures are below the melting points of the rocks (about 700°C), but they are high enough (above 250°C) for the rocks to change by recrystallization and chemical reactions.

Regional and Contact Metamorphism

Metamorphism may take place over a widespread or a limited area. Where high pressures and temperatures extend over large regions, rocks are subject to **regional metamorphism.** Regional metamorphism accompanies plate collisions that result in mountain building and the folding and breaking of sedimentary layers that once were horizontal (Figure 3.5a). Where high temperatures are restricted to smaller areas, such as the rocks near and in contact with an intrusion, rocks are transformed by **contact metamorphism** (Figure 3.5b).

Many regionally metamorphosed rocks, such as schists, have characteristic **foliation,** wavy or flat planes produced when the rock was structurally deformed into folds. Granular textures are more typical of most contact metamorphic rocks, which contain minerals with equant-shaped crystals, and of some regional metamorphic rocks formed by very high pressure and temperature.

Common Minerals

Silicates are the most abundant minerals of metamorphic rocks because these rocks are transformations of other rocks that are rich in silicates (see Table 3.1). Typical minerals of metamorphic rocks are quartz, feldspar, mica, pyroxene, and amphibole, the same kinds of silicates characteristic of igneous rocks. Several other silicates—kyanite, staurolite, and some varieties of garnet—are characteristic of metamorphic rocks alone. They form under conditions of high pressure and temperature in a crustal setting

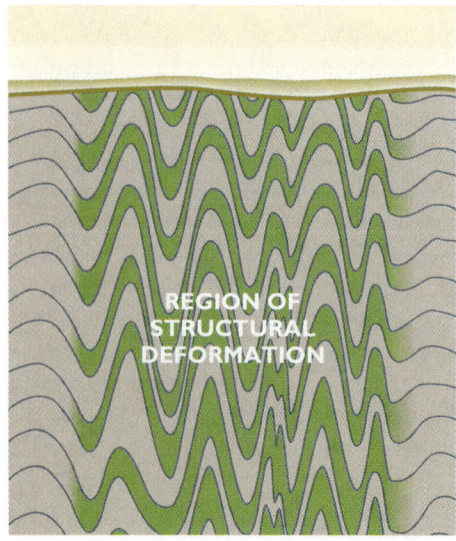

(a) Regional metamorphism

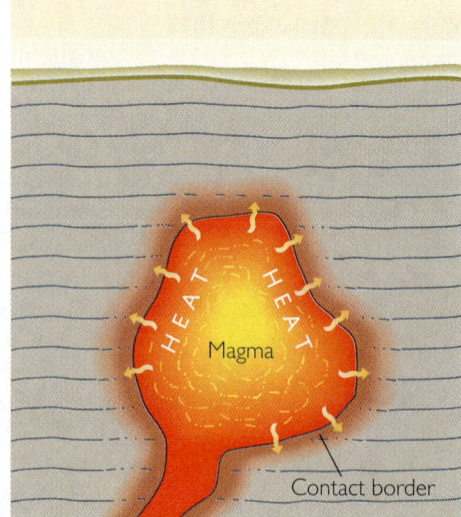

(b) Contact metamorphism

FIGURE 3.5 Metamorphism may take place over a widespread or limited area. (a) Regional metamorphism occurs where high pressures and temperatures extend over large areas, as happens when heat and pressure accompanying plate collisions strongly deform rocks. (b) Contact metamorphism occurs in more limited areas, when rocks in border zones of an intrusion of magma are metamorphosed by heat emanating from the magma.

and are not characteristic of igneous rocks. They are therefore good indicators of metamorphism. Calcite is the main mineral of marbles, which are metamorphosed limestones.

THE CHEMICAL COMPOSITION OF ROCKS

When geologists try to infer the geologic origin of rocks, they frequently make chemical analyses to determine the relative proportions of the rocks' chemical elements. Chemical analyses complement mineralogical studies, particularly for the very fine-grained or glassy rocks, such as volcanic lavas, in which few individual minerals can be observed, even with a microscope. The chemical composition of one of the abundant igneous rocks, basalt, for example, reveals the kinds of rock that melted to form the magma from which the basalt crystallized. By convention the elements are stated in terms of oxides, the compounds they form by combining with oxygen, the most abundant element in the Earth. This convention is followed even though the oxides are not actually present as such but are mostly in the form of silicates. Table 3.2 shows the major elements, as percentages by weight, for a sample of basalt. In that table you will see that the element silicon, for example, is given as the oxide silica (SiO_2) and that it makes up 48 percent of the basalt. In an ideal analysis, the percentages of all the elements would total 100, but small errors in analysis or the omission of minor elements, as in Table 3.2, may produce a smaller total.

Basalt is a typical igneous rock; about two-thirds of its weight consists of silica and alumina, the oxide of aluminum. Iron oxides are present in smaller amounts (14.7 percent), followed by calcium, magnesium, sodium, and potassium oxides (see Table 3.2). These seven *major elements,* along with oxygen, make up the great bulk of all rocks in the Earth.

So how can such an analysis help us understand basalt's geologic origin? Differences of a few percentage points or less in the proportions of some elements—such as the major cations—may indicate, for example, whether a basalt was formed at a mid-ocean ridge, where plates diverge, or at a subduction zone, where plates converge.

One difference geologists watch for in distinguishing among different kinds of rocks is the amount of water (the oxide of hydrogen) chemically bound in minerals. Water is a constituent of many rocks, but rarely in abundance. Many igneous rocks, for example, are only about 1 percent water, whereas sedimentary rocks may contain as much as 5 percent water, not counting the water present as liquid in their pore spaces. The larger amount of water in sedimentary rocks is traceable to the abundance of clay minerals in them. Finding a rock with such a large water content, a geologist would expect it to be a sedimentary rock, not an igneous rock.

In Chapter 4 we will look more closely at chemical differences among igneous rocks, and at the way these differences function as clues to the origin of the magmas from which the rocks came and the conditions under which they crystallized. In Chapter 7 we examine in more detail the chemical compositions of sedimentary rocks and the way they indicate the kinds of rocks that weathered and provided the sediment, as well as the chemical conditions under which precipitated minerals were formed. Similarly, in Chapter 8 we will see that chemical analyses of metamorphic rocks are guides to the preexisting rocks that were transformed by heat and pressure.

WHERE WE SEE ROCKS

Rocks are not found in nature conveniently divided into separate bodies—igneous here, sedimentary there, metamorphic in another place. Instead, they are found jumbled together in patterns determined

TABLE 3.2

CHEMICAL ANALYSIS OF MAJOR ELEMENTS OF BASALT, AN IGNEOUS ROCK

ELEMENT	FORMULA[1]	PERCENT OF WEIGHT
Silicon	SiO_2	48.0
Aluminum	Al_2O_3	16.0
Iron	Fe_2O_3, FeO	14.7
Calcium	CaO	10.0
Magnesium	MgO	3.9
Sodium	Na_2O	3.5
Potassium	K_2O	1.5
All major elements		97.6[2]

[1]Formulas are conventionally shown as oxides, though the elements are not actually present in the rock in this form.

[2]Percentages do not add up to 100 because of the omission of minor elements and small errors in analysis.

by the geologic history of the region. Geologists map those patterns both at the surface and projected into the interior, and try to deduce the geologic past from the present variety and distribution of the rocks.

If we were to drill a hole into any spot on Earth, we would find rocks that reflect the geologic history of that region. In the top few kilometers of most regions we would probably find sedimentary rock. Drilling deeper, perhaps 6 to 10 km down, we would eventually penetrate an underlying area of older igneous and metamorphic rock.

In fact, thousands of relatively shallow holes have been drilled on the continents in the search for oil, water, and mineral resources, and these holes are major sources of information, mainly about sedimentary rocks and their history. In the quest for more data on the deep continental crust, the governments of several countries, including the United States, Germany, and Russia, have drilled to great depths on the continents. The deepest hole, in Russia, measures more than 12 km, exceeding the depth of any commercial drilling.

A large part of our knowledge about the rocks of the ocean floor comes from the hundreds of holes punched down by the Deep-Sea Drilling Program, an ongoing project to drill the world's seafloor for geological information. Started by the United States in the late 1960s, at the same time that plate tectonics swept the geological community, it is now an international program (the Ocean Drilling Program) carried on with the cooperation of the major maritime countries of the world.

Even with all these sources of information on what lies beneath Earth's surface, geologists continue to rely on the rocks exposed in **outcrops,** places where bedrock—the underlying rock beneath the loose surface materials—is laid bare (Figure 3.6). Outcrops vary from region to region because they reflect the geologic structure of the Earth at a particular spot. On a trip across North America, we might run across many kinds of outcrops. Starting from the Pacific, we would encounter sea cliffs from Mexico to Canada. From the West Coast to the Rocky Mountain front, which stretches from New Mexico in the south to Alberta in Canada, outcrops of all kinds of rock are abundant in the canyons, mountainsides, and cliffs of the relatively dry mountainous regions of the western third of the continent (Figure 3.7).

From the Rockies eastward to the Appalachian Mountains, the landscape is dominated by the plains and prairies of the American Midwest and the Canadian Plains provinces (Figure 3.8). In this region, outcrops are scarce because most of the sedi-

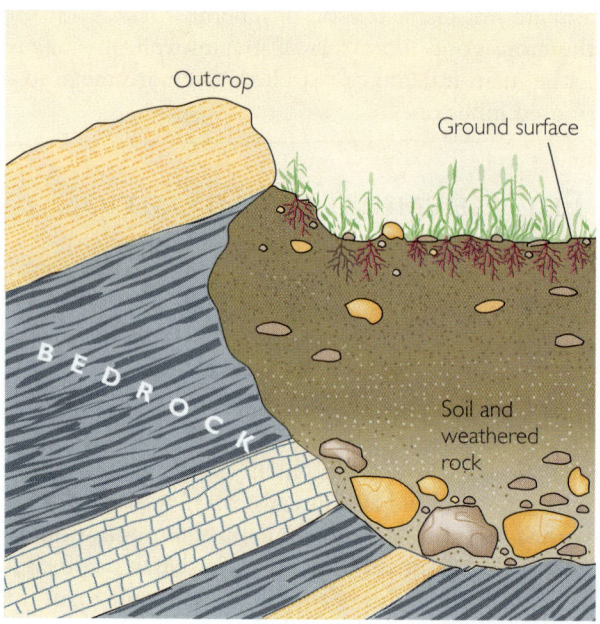

FIGURE 3.6 Outcrops are places where bedrock—the underlying rock beneath loose surface materials such as soil and boulders—is laid bare.

mentary bedrock is covered by soils and the sediments deposited by such rivers as the Missouri and their tributaries. Here geologists search the low hills and gentle valleys, looking for evidence of geologic history along dry creeks and interstate highway cuts.

Outcrops become more numerous once we reach the Appalachians. In this more humid climate, most of the rocks of the low ridges are covered by abundant vegetation and soil, but there are many outcrops along rocky cliffs and ledges, especially on the higher ridges and mountains.

Low coastal plains cover the region from southeastern New Jersey to the Carolinas and Georgia in the east and Texas, Louisiana, Mississippi, and Alabama in the south. Here barely lithified, relatively soft sedimentary rocks are exposed in outcrops similar to those of the Great Plains. Good exposures can be found in the occasional bluff along the shoreline. To the north, in the hilly, rugged landscape of New England and the Maritime provinces of Canada, we can find good outcrops, with the best exposures displayed along rocky coastlines (Figure 3.9). Other good outcrops are displayed in creek bottoms.

As this travelogue indicates, the presence and type of outcrops depend on the nature of the landscape, which in turn depends on the geologic structure of the region, its history, and its present climate. In later chapters, we explore in more detail the way rock types relate to geologic structures (Chapter 10)

FIGURE 3.7 Rocky cliffs on the Pacific coast at Cape Kiwanda, Oregon. Shoreline cliffs such as these provide ready accessibility to bedrock for the geologist. *(Fred Hirschmann.)*

FIGURE 3.8 Granite outcrops on Bonaventure Island, Quebec, Canada, form low ledges of bedrock, often the only exposure of bedrock in gentle topography. In some cases this type of outcrop extends to the Atlantic seaboard of Canada. *(Russ Kinne/Comstock.)*

FIGURE 3.9 Shawangunk Mountains, New York. Even though these are ancient mountains, part of the Appalachian Mountain chain, excellent outcrops of bedrock are formed along steep slopes. *(Carr Clifton.)*

and to landscape (Chapter 16). Now, however, we turn to the rock cycle, which—in combination with plate tectonics—reveals the interrelationships among the three groups of rocks and thus geologic structure and history.

THE ROCK CYCLE

The **rock cycle** is a set of geologic processes by which each of the three great groups of rocks is formed from the other two. The Scotsman James Hutton described this cycle in an oral presentation in 1785 before the Royal Society of Edinburgh; ten years later he presented it in more detail in his book *Theory of the Earth with Proof and Illustrations.* As is often the case in the history of science, other scientists, both in England and on the European continent, had also recognized elements of the cyclic nature of geological change. Hutton's role was that of synthesizer—he presented the larger picture that has enabled us to understand the process.

We present an account of one particular cycle here, recognizing that such cycles vary with time and place. We can start our account with a magma deep in Earth's interior, where temperatures and pressures are high enough to melt any kind of pre-existing rock: igneous, metamorphic, or sedimentary (see Figure 3.10). This activity deep in Earth's crust Hutton called the *plutonic episode,* for Pluto, the Roman god of the underworld. We now refer to all igneous intrusives as **plutonic rocks,** whereas the extrusives are known as **volcanic rocks.** As preexisting rocks melt, all their component minerals are destroyed and their chemical elements are homogenized in the resulting hot liquids. As the magma cools, crystals of new minerals grow and form new

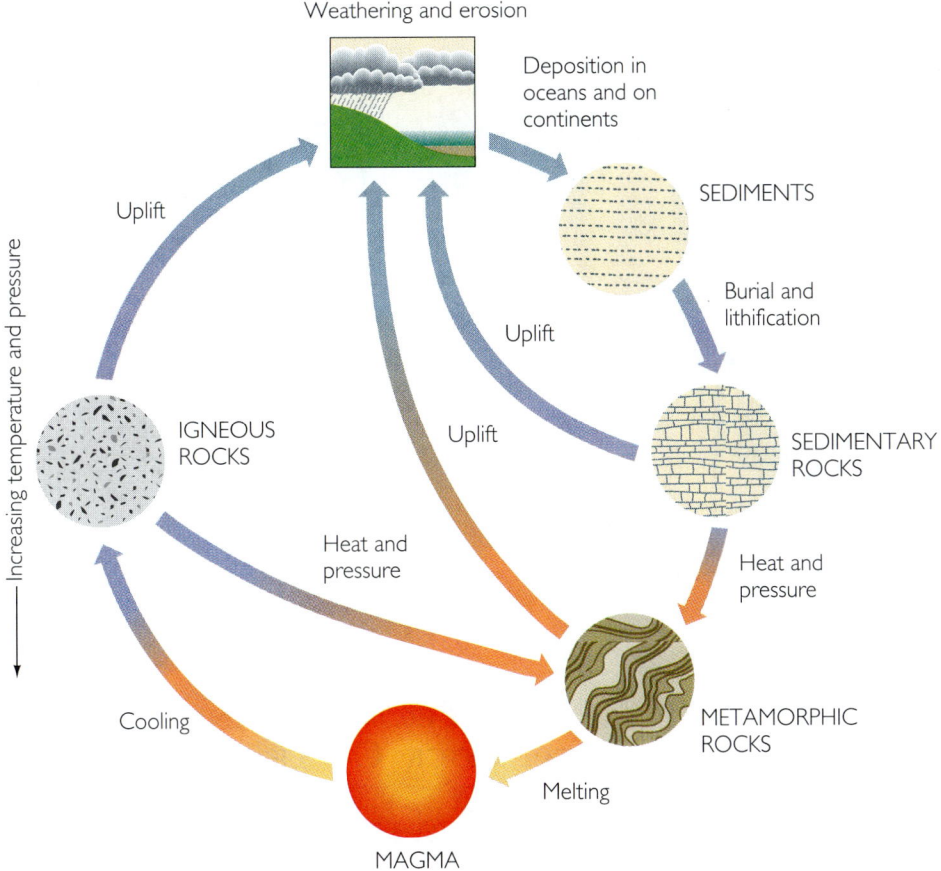

FIGURE 3.10 The rock cycle, as proposed by James Hutton 200 years ago. Subjected to weathering and erosion, rocks form sediments, which are deposited, buried, and lithified. After deep burial, the rocks undergo metamorphism, melting, or both. Through orogeny and volcanic processes, rocks are uplifted, only to be recycled again.

igneous rock. Melting and formation of igneous rock take place mainly along the boundaries of colliding or diverging tectonic plates, as well as in mantle plumes, as you will see in later chapters.

The igneous rocks that form at the boundaries where plates collide, together with associated sedimentary and metamorphic rocks, are then uplifted as a high mountain chain as a section of Earth's crust becomes crumpled and deformed. Geologists call this process, which begins with plate collision and ends in mountain building, **orogeny.** After uplift, the rocks of the crust overlying the uplifted igneous rock gradually weather, creating loose material that erosion then strips away, exposing the igneous rock at the surface.

The igneous rock, now in cool, wet surroundings far from its birthplace in the hot interior, also weathers, and some of its minerals undergo chemical changes. Iron minerals, for example, may "rust" to form iron oxides. High-temperature minerals such as feldspars may become low-temperature clay minerals. Such substances as pyroxene may dissolve completely as rain pours over them. The weathering of the igneous rock again produces various sizes and kinds of rock debris and dissolved material, which are carried away by erosion. Some are transported by water and wind on land. Much of the debris is transported by streams to rivers and ultimately to the ocean, where it is deposited as layers of sand, silt, and other sediments formed from dissolved material, such as the calcium carbonate from shells.

These sediments laid down in the sea, as well as those deposited by water or wind on the land, are buried under successive layers of sediment, where they gradually lithify into sedimentary rock. Burial is accompanied by **subsidence;** that is, a depression or sinking of the Earth's crust. As subsidence continues, additional layers of sediment can accumulate.

FIGURE 3.11 Many processes combine to produce the rock cycle. These processes are all driven by plate tectonics. The five parts of this figure illustrate those relationships in various tectonic settings.

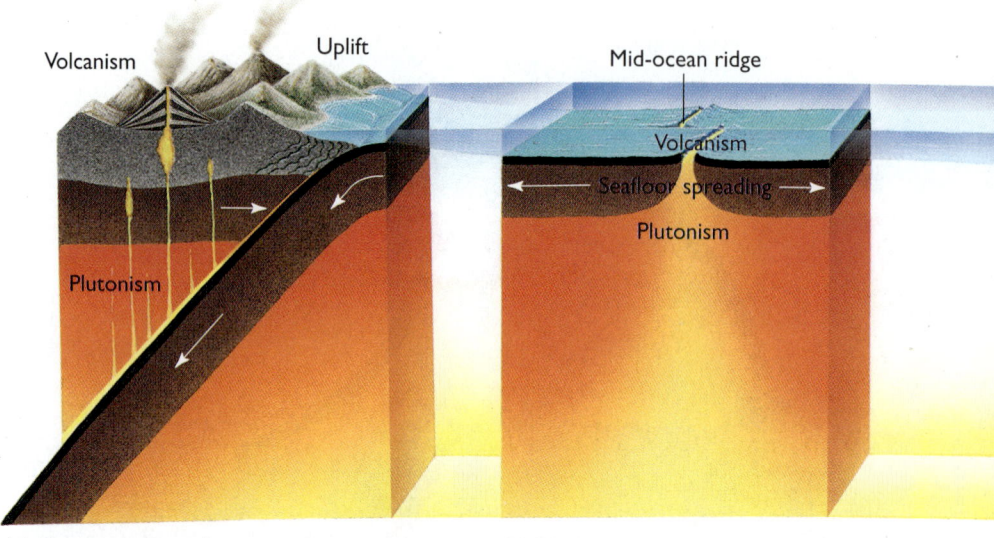

(a) Convergent boundary: oceanic plates subduct and melt.

(b) Divergent boundary: magmas rise and seafloors spread.

As the lithified sedimentary rock is buried more deeply in the crust, it gets hotter. As the depth of burial exceeds 10 km and temperatures climb to over 300°C, the minerals in the still-solid rock start to change to new minerals that are more stable at the higher temperatures and pressures of the deeper parts of the crust. This is the process of metamorphism, which transforms the previously sedimentary rocks into metamorphic rocks. With further heating, the rocks may melt and form a new magma from which igneous rocks will crystallize, starting the cycle all over again.

As we noted earlier, this cycle is only one variation among the many that may occur in the rock cycle. Any type of rock—metamorphic, sedimentary, or igneous—can be uplifted during an orogeny and weathered and eroded to form new sediments. Some stages may be omitted; as a sedimentary rock is uplifted and eroded, for example, metamorphism and melting are skipped. And stages may occur out of sequence, as when an igneous rock formed in the interior is metamorphosed before it is uplifted. And, as we know from deep drilling, some igneous rocks many kilometers down in the crust may never have been uplifted or exposed to weathering and erosion.

The rock cycle never ends; it is always operating at different stages in different parts of the world, forming and eroding mountains in one place and laying down and burying sediments in another. The rocks that make up the solid Earth are recycled continuously, but we can see only the surface parts of the cycle and must deduce the recycling of the deep crust and the mantle from indirect evidence.

PLATE TECTONICS AND THE ROCK CYCLE

Plutonism, volcanism, tectonic uplift, metamorphism, weathering, sedimentation, transportation, deposition, and burial are the geologic processes that combine in the rock cycle to convert the three groups of rocks to one another. These processes are all driven by plate tectonics.

Plutonism and volcanism are the result of the interior heat of the Earth, and we identify them primarily with three plate-tectonic settings:

- Convergent boundaries, where oceanic plates descend into the mantle and melt, and igneous rocks eventually form (Figure 3.11a)

- Divergent boundaries at mid-ocean ridges, where seafloor spreading allows basaltic magmas to rise from Earth's mantle to form oceanic crust (Figure 3.11b)

- Mantle plumes or hot spots, where magmas rise through the mantle and pour out at the surface (Figure 3.11d)

Sediment is carried away from mountains and deposited on continents and ocean floors as plates slowly subside while they move away from mid-ocean ridges. Multiple layers of sediment accumulate, burying lower layers (Figure 3.11c).

In contrast to the Earth's interior, at the surface the Sun's heat drives the circulation of the oceans

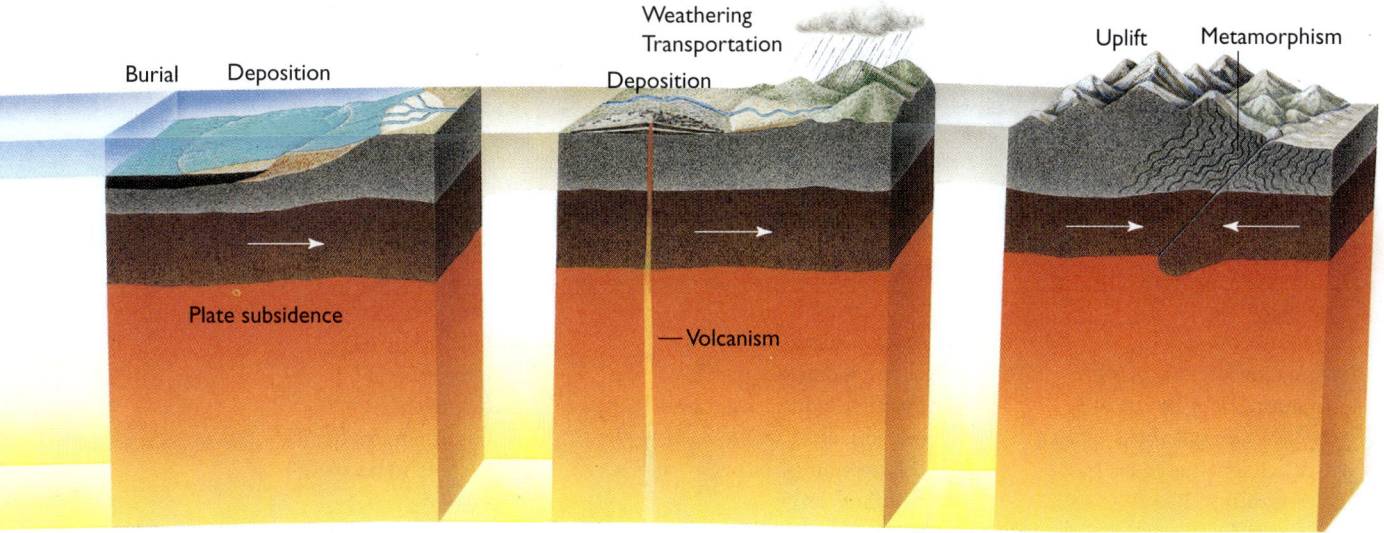

(c) Ocean floor and shoreline: subsidence of an oceanic plate causes deposition of sediment, burial, and lithification.

(d) Continental interior: stable plates are dominated by weathering, deposition, and transportation. Volcanism may be generated by mantle plumes or "hot spots."

(e) Convergent boundaries: plates carrying continents collide, uplifting mountains and metamorphosing large regions.

and atmosphere, producing weathering and transportation of sediment by wind, water, and ice (Figure 3.11d).

Metamorphism and uplift occur when continental plates collide at convergent boundaries. These collisions uplift mountains and create the great pressures and temperatures that metamorphose rocks (Figure 3.11e).

With this introduction to the rock world, we are ready to begin the study of rocks. In Chapters 4 and 5 we look at the geologic origin of magmas, the types of igneous rocks that form by their crystallization, the larger picture of plate-tectonic control of igneous processes, and the dynamics of volcanoes and their eruptions. In Chapters 6 and 7 we explore weathering, the nature of sedimentary particles, and the ways various sediments and sedimentary rocks are produced. We complete our discussion of rocks in Chapter 8 by seeing how high heat and pressure affect preexisting rocks, transforming them to metamorphic rocks, and how metamorphism relates to plate tectonics and orogeny.

Summary

What determines the properties of the various kinds of rocks that are formed in and on the surface of the Earth? Mineralogy (the kinds and proportions of minerals that make up a rock) and texture (the sizes, shapes, and spatial arrangement of its crystals or grains) define a rock. The mineralogy and texture of a rock are determined by the geologic conditions, including chemical composition, under which it formed, either in the interior under various conditions of high temperature and pressure or at the surface, where temperatures and pressures are low.

What are the three types of rock and how are they formed? Igneous rocks are formed by the crystallization of magmas as they cool. Intrusive igneous rocks form in Earth's interior and have large crystals. Extrusive igneous rocks, which form at the surface of the Earth where lavas and ash erupt from volcanoes, have a glassy or fine-grained texture. Sedimentary rocks are formed by the lithification of sediments after burial. Sediments are derived from the weathering and erosion of rocks exposed at the Earth's surface. Metamorphic rocks are formed by alteration in the solid state of igneous, sedimentary, or other metamorphic rocks as they are subjected to high temperatures and pressures in the interior.

How does the rock cycle describe the formation of rocks as the products of geologic processes? The rock cycle relates rock-forming geologic processes to the formation of the three types of rocks from one another. One can view the processes by starting at any point in the cycle. We began with the formation of igneous rocks by crystallization of a magma in the interior of the Earth. The igneous rock then is uplifted to the surface in the mountain-building process. There it is exposed to weathering and erosion, which produce sediment. The sediment is cycled back to the interior by burial and lithification to sedimentary rock. Deep burial leads to metamorphism or melting, at which point the cycle begins again. Plate tectonics is the mechanism by which the cycle operates.

Key Terms and Concepts

texture (p. 60)
igneous rocks (p. 60)
sedimentary rocks (p. 60)
metamorphic rocks (p. 60)
intrusive igneous rocks (p. 61)
extrusive igneous rocks (p. 62)
volcanism (p. 62)
sediments (p. 62)

weathering (p. 62)
erosion (p. 62)
clastic sediments (p. 62)
chemical and biochemical
 sediments (p. 62)
lithification (p. 62)
bedding (p. 62)
regional metamorphism (p. 64)

contact metamorphism (p. 64)
foliation (p. 64)
outcrops (p. 66)
rock cycle (p. 68)
plutonic rocks (p. 68)
volcanic rocks (p. 68)
orogeny (p. 69)
subsidence (p. 69)

Exercises

1. What are the differences between extrusive and intrusive igneous rocks?

2. What are the differences between regional and contact metamorphism?

3. What are the differences between clastic and chemical or biochemical sedimentary rocks?

4. List three common silicate minerals found in each group of rocks: igneous, sedimentary, and metamorphic.

5. Of the three groups of rocks, which are formed at Earth's surface and which in the interior of the crust?

6. Where on the continents can you see bare rock?

Thought Questions

1. What geologic processes transform a sedimentary rock to an igneous rock?

2. Name a mineral found only in sedimentary rocks that you might use to distinguish between a fine-grained sedimentary rock formed from lithified mud and an extrusive igneous rock.

3. As a magma cools, what might cause differences in the sizes of the crystals of two intrusive igneous rocks, one with crystals about 1 cm in diameter, the other with crystals about 2 mm in diameter?

4. Which igneous intrusion would you expect to have a wider contact metamorphic zone: one intruded by a very hot magma or one intruded by a magma of a more moderate temperature?

5. Describe the geologic processes by which an igneous rock is transformed to a metamorphic rock and then exposed to erosion.

6. Describe the kinds of outcrop that are found in various places in your hometown. If none, explain how you would determine the nature of the buried bedrock.

7. How does plate tectonics explain plutonism?

SHORT-TERM TEAM PROJECT: IDENTIFYING BUILDING STONES

Building stones often are a clue to local geology. Local stone is both cost-effective and a source of community pride. With a partner, examine the building stones you find on campus or in your community. Choose between four and six different-looking types of stone and note their locations on a local map. Draw each stone on a separate piece of paper and note such features as color, grain size, presence or absence of layering, and whether the stone appears to contain one mineral or more. Also describe any evidence of chemical or physical weathering and judge how good the stone is for building. Then decide whether the stone is most likely to be igneous, metamorphic, or sedimentary, and explain why. Finally, compare a geologic map of the area with your findings in the field and explain why the stones you described do or do not reflect local geology. Submit an organized folder containing your drawings, observations, and inferences.

SUGGESTED READINGS

Blatt, Harvey, and Robert J. Tracy. 1996. *Petrology: Igneous, Sedimentary, and Metamorphic,* 2nd ed. New York: W. H. Freeman.

Dietrich, Robert V., and Brian J. Skinner. 1980. *Rocks and Rock Minerals.* New York: Wiley.

Ernst, W. Gary. 1969. *Earth Materials.* Englewood Cliffs, N.J.: Prentice Hall.

Prinz, Martin, George Harlow, and J. Peters. 1978. *Simon & Schuster's Guide to Rocks and Minerals.* New York: Simon & Schuster.

INTERNET SOURCES

Rock Cycle
ⓘ http://www.science.ubc.ca/~geol202/s/rock_cycle/rockcycle.html
The University of British Columbia established a Web site for its Introduction to Petrology course (Geology 202) with a variety of URLs. This URL will take you directly to the Rock Cycle.

Rock Families
ⓘ http://www.science.ubc.ca/~geol202/
This URL for the University of British Columbia's Introduction to Petrology course (Geology 202) provides links to a variety of data and images concerning igneous, sedimentary, and metamorphic rocks. There are links to additional pages on terminology and classification of rocks and an Image Gallery that can be searched.

4

Columnar basalts, North Iceland. Masses of this kind of extrusive igneous rock fracture along more or less symmetrical columnar joints when they cool. (*Thomas Dressler/DRK.*)

Igneous Rocks: Solids from Melts

Igneous rocks are formed by the solidification of magmas. Chapter 3 examined igneous rocks in the context of the rock cycle; Chapter 5 discusses them in the context of the magmas that feed volcanoes. In this chapter, we take a broader and yet more detailed view of the wide range of igneous rocks, both intrusive and extrusive, and the processes by which they form.

Geologists' understanding of the variety and formation of igneous rocks is a rich chapter in the history of science. It began more than 2000 years ago, when the Greek scientist and geographer Strabo traveled to Sicily to observe the volcanic eruptions of Mount Etna. Strabo observed that the hot liquid lava spilling down from this active volcano onto Earth's surface cooled and solidified into solid rock within a few hours. Two millennia later, eighteenth-century geologists made another connection, and it would be a milestone in the development of geology into a modern observational and experimental science. They began to understand that some of the bands or sheets of rock that cut across other formations had also been formed by the cooling and solidifying of magma. In this case, however, the magma had cooled and solidified slowly because it had remained buried in the Earth's crust. Today we know that at places deep in the hot crust and mantle of the Earth, rocks melt and rise toward the surface. Some solidify before they reach the surface, and some break through and solidify on the surface. Both groups are igneous rocks.

As we saw in Chapter 3, much of Earth's crust is composed of igneous rock, some metamorphosed and some not. Since igneous rocks are the products of the cooling and solidification of liquid rock, it follows that understanding the processes that melt and resolidify rocks is a key to understanding how Earth's crust is formed.

As we also saw in Chapter 3, modern geologists have learned that the origins of a wide variety of igneous rocks are connected with plate tectonics, especially with the spreading apart of two divergent plates and the sinking of one convergent plate below another. Although much is still to be learned about the exact *mechanisms* of melting and solidification, geologists now have good answers to questions that long perplexed them: How do igneous rocks differ from one another? Where do igneous rocks form? How do rocks solidify from a melt? Where do melts form?

How Do Igneous Rocks Differ From One Another?

Late eighteenth-century geologists had sought an understanding of igneous rocks in their field observations. By the late nineteenth century, some geologists had moved their inquiries to the laboratory, running experiments to determine the mineral and chemical composition of igneous rocks collected in the field. They classified their rock samples in the same general way we do today:

- By texture.
- By mineral and chemical composition.

Texture

Two hundred years ago, the first division of igneous rocks was made on the basis of texture, as geologists classified rocks as either coarsely or finely crystalline. Texture is a simple and practical distinction that a geologist can easily see in the field. A coarse-grained rock such as granite has separate crystals quite visible to the naked eye. In contrast, the crystals of fine-grained rocks such as basalt are too small to be seen, even with the aid of the small magnifying lens that no field geologist would be without. (Figure 4.1 shows two samples of granite and basalt, accompanied by photographs, called photomicrographs, of very thin transparent rock slices. Photomicrographs, which are taken through a microscope, give an enlarged view of minerals and their textures.) The textural difference was clear to early geologists; its meaning would be unraveled only through further work.

First Clue: Volcanic Rocks The first clue to the meaning of texture came from the study of volcanic rocks. Early geologists could observe rocks forming from lava during volcanic eruptions. (*Lava*, you may recall from Chapter 3, is the term we apply to magma flowing out on the surface.) Geologists noted that when lava cooled rapidly, it formed either a finely crystalline rock or a glassy one in which no crystals could be distinguished. Where lava cooled more slowly, as in the middle of a thick flow many meters high, somewhat larger crystals were present.

Second Clue: Laboratory Studies of Crystallization The second clue to the implications of texture came in the nineteenth century, as experimental scientists studying familiar liquids came to understand the nature of crystallization. Anyone who has frozen an ice cube knows that water solidifies to ice in a few hours as its temperature drops below the freezing point. If you have ever attempted to retrieve your ice cubes before they were completely solid, you may have seen thin ice crystals forming at the surface and along the sides of the tray. During crystallization, the water molecules take up fixed positions in the solidifying crystal structure, and they are no longer able to move freely, as they did when the water was liquid. All other liquids, including magmas, crystallize in this way.

The first tiny crystals form a pattern. Other atoms or ions in the crystallizing liquid then attach themselves in such a way that the tiny crystals grow larger. It takes some time for the atoms or ions to "find" their correct places on a growing crystal, and large crystals form only if they have time to grow slowly. Recall from Chapter 3 that if a liquid solidifies very quickly, as a magma does when it erupts onto the cool surface of the Earth, the crystals have no time to grow larger. Instead, a large number of tiny crystals form simultaneously as the liquid cools and solidifies.

FIGURE 4.1 Texture was the first characteristic by which igneous rocks were classified. Early geologists assessed texture with a small hand-held magnifying lens. Modern geologists have access to high-powered polarizing microscopes, which produce photomicrographs of thin, transparent rock slices like those shown here. The coarsely crystalline rock on the left is granite; the finely crystalline rock on the right is basalt. *(Hand-sample photos by Chip Clark. Photomicrographs by Raymond Siever.)*

THIRD CLUE: GRANITE—EVIDENCE OF SLOW COOLING The study of volcanoes allowed early geologists to link finely crystalline textures with quick cooling at Earth's surface and to see finely crystalline igneous rocks as evidence of former volcanism. But in the absence of direct observation, what geological evidence could allow them to deduce that coarse-grained rocks formed by slow cooling deep in the interior? Granite—one of the commonest rocks of the continents—turned out to be the crucial clue (Figure 4.2). James Hutton, one of the founders of modern geology, saw granites cutting across and disrupting the bedding, or layers, of sedimentary rocks as he worked in the field in Scotland near the end of the eighteenth century. He noticed that the granite had somehow fractured and invaded the sedimentary rocks, as though the granite had been forced into the fractures as a liquid.

As Hutton looked at more and more granites, he began to focus on the sedimentary rocks bordering them. He observed that the minerals of the sedimentary rocks in contact with the granite were

FIGURE 4.2 Granite exposed in Joshua Tree National Monument, California. Weathering has brought out the coarsely crystalline texture of this intrusive igneous rock, which is made up of large crystals of quartz, feldspar, and other silicates. *(David Muench.)*

different from those found in sedimentary rocks at some distance from the granite. He concluded that the changes in the sedimentary rocks must have resulted from great heat and that the heat must have emanated from the granite. Hutton also noted that granite was composed of crystals interlocked like the pieces of a jigsaw puzzle (see Figure 4.1). By this time, chemists had established that a slow crystallization process produces this pattern.

Assessing these three lines of evidence, Hutton proposed that granite had formed from a hot molten material that solidified deep in the Earth. The evidence was conclusive, as no other explanation could accommodate all the facts. Other geologists, seeing the same characteristics of granites in widely separated places in the world, came to recognize that granite and many similar coarsely crystalline rocks were the products of magma that had crystallized slowly in the interior of the Earth.

INTRUSIVE IGNEOUS ROCKS The full significance of textural distinction in igneous rocks is now clear. As we have seen, texture is linked to the rapidity, and therefore the place, of cooling. Slow cooling of magma in Earth's interior allows adequate time for the growth of the interlocking large coarse crystals that characterize intrusive igneous rocks, also known as plutonic rocks. An **intrusive igneous rock** is igneous rock that has forced its way into surrounding rock. This surrounding rock is called **country rock.** (Later in this chapter we examine some special forms of intrusive igneous rocks.)

EXTRUSIVE IGNEOUS ROCKS Rapid cooling at Earth's surface produces the finely grained texture or glassy appearance of the **extrusive igneous rocks.** These rocks form when lava or other volcanic material erupts from volcanoes; therefore, they are also known as volcanic rocks. They are of two major categories:

- *Lavas* Volcanic rocks formed from lavas range in appearance from smooth and ropy to sharp, spiky, and jagged. We now know that many of these special textural properties are clues to the conditions under which these rocks formed, the composition of their parent magma, and the way they were ejected from the volcano.

- *Pyroclastic rocks* In more violent eruptions, pyroclasts are formed when broken pieces of lava and glass are thrown high in the air (Figure 4.3). **Pyroclasts** include crystals that started to form before the explosion, fragments of previously solidified lava, and pieces of glass that cooled and then fractured during the eruption. The finest pyroclasts are **volcanic ash,** extremely small fragments, usually of glass, that form when escaping gases force a fine spray of magma from a volcano. Volcanic ash accumulates as layers of loose and uncemented material. All volcanic rocks lithified from these pyroclastic materials are called **tuff.**

Volcanic glass comes in a variety of forms when it is the sole constituent of an igneous rock. One common glassy rock type is **pumice,** a frothy mass with a great number of *vesicles*, holes that remained after trapped gas escaped from the solidifying melt. Another wholly glassy volcanic rock is **obsidian,** which, unlike pumice, contained no trapped gases and so is solid and dense. Chipped or fragmented obsidian produces very sharp edges, and Native

FIGURE 4.3 Pyroclasts form when violent volcanic eruptions hurl broken pieces of lava and glass high in the air. Obsidian, shown here on the left, is a solid glassy pyroclast. Volcanic ash, in the center, forms when a fine spray of magma erupts. The multiple small holes in pumice, on the right, are evidence of the gas formerly trapped in this pyroclast. *(Chip Clark.)*

FIGURE 4.4 A quartz-rich felsic porphyry. The white larger crystals are phenocrysts of feldspar. *(A. J. Copley/ Visuals Unlimited.)*

Americans and many other hunting groups used it for arrowheads and a variety of cutting tools.

A **porphyry** has a mixed texture in which large crystals "float" in a predominantly fine crystalline matrix (Figure 4.4) The large crystals, called *phenocrysts*, formed while the magma was still below Earth's surface. Then, before other crystals could grow, a volcanic eruption brought the magma to Earth's surface, where as lava it quickly cooled to a finely crystalline mass. In Chapter 5 we will look more closely at how these and other volcanic rocks form during volcanism. Now, however, we turn to the second way igneous rocks are classified.

Chemical and Mineral Composition

Igneous rocks can be subdivided on the basis of their chemical and mineral compositions as well as according to texture. Volcanic glass, which is formless even under the microscope, is frequently classified by chemical analysis, as are some very finely grained rocks (see Chapter 3). One of the earliest classifications of igneous rocks was based on a simple chemical analysis of their silica (SiO_2) content. Silica, as noted in Chapter 3, is abundant in most igneous rocks and accounts for 40 to 70 percent of their total weight. We still refer to rocks rich in silica, such as granite, as silicic.

Modern classifications now group igneous rocks according to their relative proportions of silicate minerals (Table 4.1). These proportions were

TABLE 4.1

Common Minerals of Igneous Rocks

COMPOSITIONAL GROUP	MINERAL	CHEMICAL COMPOSITION	SILICATE STRUCTURE
FELSIC	Quartz	SiO_2	Frameworks
	Potassium feldspar	$KAlSi_3O_8$	
	Plagioclase feldspar	$\begin{cases} NaAlSi_3O_8 \\ CaAl_2Si_2O_8 \end{cases}$	
	Muscovite (mica)	$KAl_3Si_3O_{10}(OH)_2$	Sheets
MAFIC	Biotite (mica)	$\begin{rcases} K \\ Mg \\ Fe \\ Al \end{rcases} Si_3O_{10}(OH)_2$	
	Amphibole group	$\begin{rcases} Mg \\ Fe \\ Ca \\ Na \end{rcases} Si_8O_{22}(OH)_2$	Double chains
	Pyroxene group	$\begin{rcases} Mg \\ Fe \\ Ca \\ Al \end{rcases} SiO_3$	Single chains
	Olivine	$(Mg,Fe)_2SiO_4$	Isolated tetrahedra

determined by thousands of mineralogical analyses performed by geologists all over the world. The silicate minerals—quartz, feldspar (both orthoclase and plagioclase), muscovite and biotite micas, the amphibole and pyroxene groups, and olivine—form a systematic series. Felsic minerals are high in silica; mafic minerals are low in silica. The adjectives *felsic* (from *fel*dspar and *si*lica) and *mafic* (from *ma*gnesium and *fer*ric, from Latin *ferrum*, iron) are applied to both the minerals and the rocks that are high in these minerals. Mafic minerals crystallize at higher temperatures—that is, earlier in the cooling of a magma—than those at which felsic minerals crystallize.

As the mineral and chemical compositions of igneous rocks became known, geologists soon noticed that some extrusive and intrusive rocks were identical in composition and differed only in texture. Basalt, for example, is an extrusive rock formed from lava. Gabbro has exactly the same mineral composition as basalt but is formed deep in the Earth's crust (Figure 4.5). Similarly, rhyolite and granite are identical in composition and different in texture. Thus extrusive and intrusive rocks form two chemically and mineralogically parallel sets of igneous rocks. Conversely, most chemical and mineral compositions can appear as either extrusive or intrusive rocks. The sole exceptions are very highly mafic rocks that rarely or never appear as extrusive igneous rocks.

Figure 4.6 is a model that portrays the relationships we have been discussing. Notice that the horizontal axis plots silica content as a percentage of a given rock's weight. The percentages given—from high-silica content at 70 percent to low-silica content at 40 percent—cover the range found in igneous rocks. The vertical axis displays a scale measuring the mineral content of a given rock. If, for example, you had established that the silica content of a coarsely grained rock sample was about 70 percent, you could run a vertical line up from the 70 percent marker on the horizontal axis, which in fact corresponds to the left border of Figure 4.6. As the vertical line moves upward, it intersects the minerals amphibole, biotite, muscovite, plagioclase feldspar, quartz, and orthoclase feldspar, but it excludes the minerals pyroxene and olivine. Using the vertical axis to estimate the percentages of minerals at the points where the line intersects, you could then assert that the sample was probably granite, which has a mineral content like the one mapped by your vertical line. It contains about 6 percent amphibole, 3 percent biotite, 5 percent muscovite, 14 percent plagioclase feldspar, 22 percent quartz, and 50 percent orthoclase feldspar. Rhyolite, which has a mineral composition identical to granite, has the same mineral profile, but its texture would eliminate it—as a volcanic rock, rhyolite has a fine texture, not a coarse one.

We can use Figure 4.6 to help with the discussion of the intrusive and extrusive igneous rocks. We begin with the felsic rocks, on the extreme left of the model.

FIGURE 4.5 Two parallel sets of extrusive and intrusive igneous rocks. The finely grained basalt *(upper left)* is identical in composition to the coarsely grained gabbro *(upper right)*. Similarly, the finely grained rhyolite *(lower left)* is identical in composition to the coarsely grained granite *(lower right)*. *(Chip Clark.)*

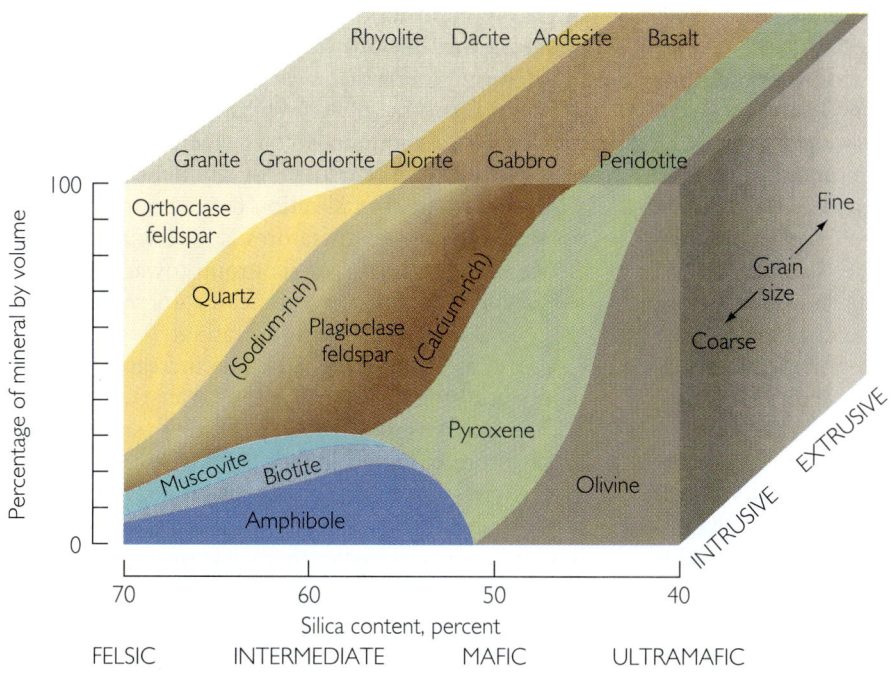

FIGURE 4.6 Classification model of igneous rocks. The vertical axis measures the mineral composition of a given rock as a percentage of its volume. The horizontal axis is a scale of silica content by weight. Thus if you knew by chemical analysis that a coarsely textured rock sample contained about 70 percent silica, you could determine that it also contained about 6 percent amphibole, 3 percent biotite, 5 percent muscovite, 14 percent plagioclase feldspar, 22 percent quartz, and 50 percent orthoclase feldspar. Your rock would be granite. Although rhyolite has the same mineral profile, its fine texture would eliminate it.

Felsic Rocks **Felsic rocks** are light-colored igneous rocks that are poor in iron and magnesium and rich in minerals that are high in silica, such as quartz, orthoclase feldspar, and plagioclase feldspar. The plagioclase feldspars contain both calcium and sodium. As Figure 4.6 indicates, they are richer in sodium near the felsic end and richer in calcium near the mafic end. In accord with the point made earlier, that mafic minerals crystallize at temperatures higher than those at which felsic minerals crystallize, calcium-rich plagioclases crystallize at higher temperatures than the sodium-rich plagioclases.

Felsic minerals and rocks tend to be light in color. **Granite,** the best known and one of the most abundant intrusive igneous rocks, is a felsic rock, with about 70 percent silica. Granite contains abundant quartz and orthoclase feldspar and a lesser amount of plagioclase feldspar (see the left side of Figure 4.6). These light-colored felsic minerals give granite its pink or gray color. It also contains small amounts of muscovite and biotite micas and amphibole.

Rhyolite is the extrusive equivalent of granite. This light-brown to gray igneous rock shares granite's felsic composition and light coloration, but it is much more finely grained. Many rhyolites are composed largely or entirely of volcanic glass.

Intermediate Igneous Rocks Midway between the felsic and mafic ends of the series are the **intermediate igneous rocks;** as their name indicates, these rocks are neither as rich in silica as the felsic rocks nor as poor in it as the mafic rocks. We find the intrusive intermediate rocks to the right of granite in Figure 4.6. The first is **granodiorite,** a light-colored felsic rock that looks something like granite. It is similar to granite in having abundant quartz, but its predominant feldspar is plagioclase, not orthoclase. To its right is **diorite,** still lower in silica and dominated by plagioclase feldspar, with little or no quartz. Diorites contain a moderate amount of the mafic minerals biotite, amphibole, and pyroxene, and they tend to be darker than granite or granodiorite.

The volcanic equivalent of granodiorite is **dacite.** To its right in the extrusive series is **andesite,** the volcanic equivalent of diorite. Andesite derives its name from the Andes, the volcanic mountain chain in South America.

Mafic Rocks **Mafic rocks** are high in pyroxenes and olivines, minerals relatively poor in silica but rich in magnesium and iron, from which these minerals get their characteristic dark colors. Here, at even lower levels of silica than those found in the intermediate range, we find **gabbro,** a coarsely grained, dark-gray intrusive igneous rock. Gabbro has an abundance of mafic minerals, especially pyroxene; it contains no quartz and only moderate amounts of calcium-rich plagioclase feldspar.

Basalt is dark gray to black and is the fine-grained extrusive equivalent of gabbro. Basalt is the most abundant igneous rock of the crust and underlies virtually all the floors of the oceans. On the continents, thick and extensive sheets of basalt make up large plateaus in some places, such as the Columbia River plateau of Washington State and the remarkable formation known as the Giant's Causeway in Northern Ireland.

ULTRAMAFIC ROCKS **Ultramafic rocks** consist primarily of mafic minerals and contain less than 10 percent feldspar. Here, at the very low silica level of only about 45 percent, we find **peridotite,** a coarsely grained, dark greenish-gray rock made up primarily of olivine with small amounts of pyroxene and amphibole. Ultramafic rocks are rarely found as extrusives. They form by the accumulation of crystals from a magma and have never constituted a liquid, hence do not form lavas.

Thus igneous rocks can be classified on the basis of their composition as well as by their texture. The compositional groups can be explained geologically by a strong correlation between mineralogy and the temperatures of crystallization. As Table 4.2 indicates, mafic minerals crystallize at temperatures higher than those at which felsic minerals crystallize. This increase in crystallization temperatures is a mirror image of the temperatures at which rocks melt. As we move from the mafic group toward the felsic rocks, silica content also increases. Increasing silica content is expressed as an increasingly complex silicate structure (see Table 4.1). The increasingly complex silicate structure forms an inverse correlation with a melted rock's ability to flow: as structure grows more complex, ability to flow decreases. Thus **viscosity**—the measure of a liquid's *resistance* to flow—increases as silica content increases. Viscosity is an important factor in the behavior of lavas, as you will see in Chapter 5.

It is clear that knowing a rock's minerals can yield important information about the conditions under which the rock's parent magma formed and crystallized. To interpret this information accurately, however, we must understand more about igneous processes, the topic we turn to next.

TABLE 4.2

CHANGES IN SOME MAJOR CHEMICAL ELEMENTS FROM FELSIC TO MAFIC ROCKS

	FELSIC	INTERMEDIATE		MAFIC
COARSE-GRAINED (INTRUSIVE)	Granite	Granodiorite	Diorite	Gabbro
FINE-GRAINED (EXTRUSIVE)	Rhyolite	Dacite	Andesite	Basalt

← SILICA INCREASING

← SODIUM INCREASING

← POTASSIUM INCREASING

CALCIUM INCREASING →

MAGNESIUM INCREASING →

IRON INCREASING →

← (Viscosity increasing)

(Melting temperature increasing) →

How Do Magmas Form?

We know from the way the Earth transmits earthquake waves that the bulk of the Earth is solid for thousands of kilometers down to the boundary of the core. The evidence of volcanic eruptions tells us, however, that there must also be liquid regions where magmas originate. How do we resolve this apparent contradiction? The answer lies in the processes that melt rocks and create magmas.

Although geologists can observe volcanic eruptions of lavas and study pyroclasts in their laboratories, most igneous processes cannot be directly observed. The study of magmas and igneous processes depends primarily on geological inferences and experimental simulations. To know where rocks melt in the Earth, for example, we have to know both the conditions under which various rocks melt and the regions of the Earth in which those conditions are found.

How Do Rocks Melt?

Although we do not yet understand the exact mechanisms of melting and solidification, geologists have learned a great deal from laboratory experiments on how rocks melt. From these experiments, we know that a rock's melting point depends on the rock's composition and on conditions of temperature and pressure.

TEMPERATURE AND MELTING As early twentieth-century geologists ran experiments on rocks, they discovered that a rock of varied composition does not melt completely at any given temperature. The **partial melting** that these early geologists had discovered occurs because the minerals that compose a rock melt at different temperatures. As temperatures rise, some minerals melt and others remain solid. If melting is halted and conditions are maintained at any given temperature, melting ceases and the current mixture of solid rock and melt is maintained. The fraction of rock that has melted at a given temperature is a **partial melt.** To visualize a partial melt, you might think of how a chocolate chip cookie would look if you heated it to the point where the chocolate chips had melted while the main part of the cookie stayed solid.

The ratio of liquid to solid in a partial melt depends on the composition and melting temperatures of the original rocks and on the temperature at the depth in the crust or mantle where melting takes place. At the lower end of its melting range, a partial melt might be less than 1 percent of the volume of the original rock. Much of the hot rock would still be solid, but appreciable amounts of liquid would be present as small droplets in the tiny spaces between crystals throughout the mass. Many partial melts of basaltic magma in the upper mantle, for example, are estimated to be only 1 percent to 2 percent melt. At the high end of the melting temperature range, much of the rock would be liquid, with lesser amounts of unmelted crystals in it. This would be the case for a reservoir of a granitic magma and crystals just below a volcano.

Geologists seized on the new knowledge of partial melts to help them determine how different kinds of magma form at different temperatures and in different regions in Earth's interior. As you can imagine, the composition of a partial melt where only the minerals with the lowest melting points have melted may be significantly different from the composition of a completely melted rock. Thus basaltic magmas that formed in different regions in the mantle may have somewhat different compositions. From this observation we could deduce that the different magmas come from different proportions of partial melt.

PRESSURE AND MELTING To get the whole story on melting, we have to consider pressure, which increases with depth in the Earth as a result of the increased weight of overlying rock. Geologists found that as they melted rocks under various pressures, increasing pressure led to increasing melting temperatures. Thus rocks at melting temperatures at Earth's surface would remain solid at the same temperature in the interior, where pressures are high. For example, a rock that melted at 1000°C at Earth's surface might have a melting temperature much higher, perhaps 1300°C, at depths in the interior where pressures are many thousands of times greater than those at the surface. It is the pressure effect that explains why rocks in most of the crust and mantle do not melt. Only where composition and both temperature and pressure are right can rock melt. You will learn more about pressure and its effects on rocks in the Earth's interior in Chapter 8, "Metamorphic Rocks."

WATER AND MELTING The many experiments on melting temperatures and partial melting paid other dividends as well. One of these was a better understanding of the role of water in rock melting. Geologists knew from analyses of natural lavas that there was water in some magmas, so they added small amounts of water to the rocks they were melting. In doing so, they discovered that the compositions of partial and complete melts vary not only with temperature and pressure but also with the amount of water present. Consider, for example, the effect of water content on pure albite, the high-sodium plagioclase feldspar, at the low pressures of the Earth's surface.

If only a small amount of water is present, pure albite will remain solid at temperatures just over 1000°C, hundreds of degrees above the boiling point of water. At these high temperatures, the water in the albite is present as a vapor (gas). If large amounts of water are present, the melting temperature of the albite will be lower, dropping to as low as 800°C (see Figure 4.7). This behavior follows the general rule that dissolving some of one substance in another lowers the melting point of the solution. If you live in a cold climate, you are probably familiar with this process because you know that towns and municipalities sprinkle salt on icy roads to lower the melting point of the ice.

In the same way, the melting temperature of the albite—and of all the feldspars and other silicate minerals—drops considerably in the presence of large amounts of water. In this case, the melting points of various silicates are lowered in proportion to the amount of water dissolved in the molten silicate. This is an important point in our knowledge of melting rocks. Water content is an important factor in lowering the melting temperatures of mixtures of sedimentary and other rocks. Sedimentary rocks contain an especially high volume of water in their pore spaces, higher than that found in igneous or metamorphic rocks. As you will see later in this chapter, this water plays an important role in melting in Earth's interior.

The Formation of Magma Chambers

Most substances have a lower density in the liquid form than in the solid form. The density of a melted rock is lower than the density of a solid rock of the same composition—that is, a given volume of melt would weigh less than the same volume of solid rock. Geologists reasoned that large bodies of magma could form in the following way. If the less dense melt were given a chance to move, it would move upward, just as oil, which is less dense than water, rises to the surface of a mixture of oil and water. Being liquid, the partial melt could move slowly upward through pores and along the boundaries between crystals of the overlying rocks. As the hot drops moved upward, they would coalesce with other drops, gradually forming larger pools of molten rock within Earth's solid interior.

We now know that the large pools of molten rock envisioned by early geologists form **magma chambers**—magma-filled cavities in the lithosphere that form as ascending drops of melted rock push aside surrounding solid rock. Magma chambers may encompass a volume as large as several cubic kilometers. The exact manner in which they form is still the subject of ongoing research, and we cannot yet say exactly what magma chambers look like in three dimensions. We think of them as large liquid-filled cavities in solid rock, which expand as more of the surrounding rock melts or as liquid migrates in through cracks and other small openings between crystals. Magma chambers contract as they expel magma to the surface in eruptions. We know magma chambers exist because earthquake waves can show us the depth, size, and general outlines of magma chambers underlying some active volcanoes.

With this knowledge of how rocks could melt to form magma, we can better consider where different kinds of magma form at different temperatures and regions in Earth's interior.

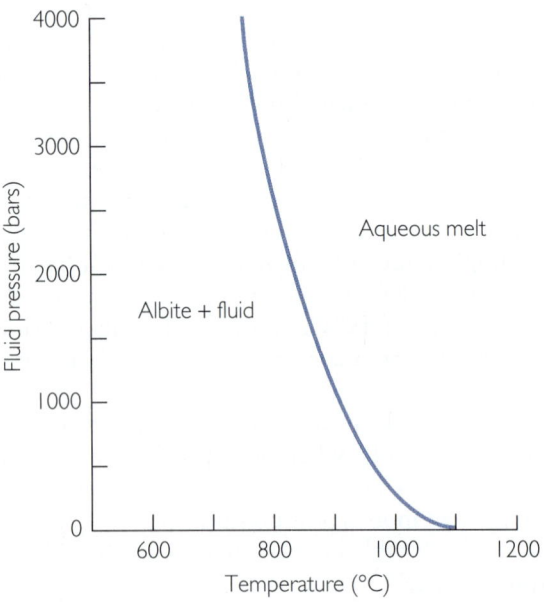

FIGURE 4.7 The melting curve for albite in the presence of water. Albite is a silicate mineral, and the melting temperature of silicate minerals drops considerably in the presence of large amounts of water. When only a small amount of water is present, pure albite remains solid at temperatures just over 1000°C; when large amounts of water are present, albite's melting point may drop as low as 800°C.

Where Does Magma Form?

We noted earlier in this chapter that our understanding of igneous processes stems from geological inferences as well as laboratory experimentation. Our inferences are based mainly on data from two sources. The first is volcanoes on land and under the sea—everywhere that molten rock erupts—which give us information about where magmas are located. The second source of data is temperatures recorded at deep drill holes and mine shafts in various locations, which show that the temperature of the interior of the Earth increases with depth. Using such figures, scientists have been able to estimate the rate at which temperature rises as depth increases.

Some temperatures recorded at various locations are much hotter than others at the same depth, an indication that some parts of Earth's mantle and crust are hotter than others. In tectonically and volcanically active areas, for example, temperature increases at an exceptionally rapid rate, reaching 1000°C at a depth of 40 km, not far below the base of the crust. This temperature is almost high enough to melt basalt. In tectonically stable regions, temperature rises much more slowly, reaching only 500°C at the same depth.

We now know that different kinds of rock can solidify from a magma through the process of partial melting. And we know that increasing temperatures in Earth's interior could cause magmas to form. But our picture is not complete; for more information, we turn again to the theory of plate tectonics.

Tectonic Activity, Rock Composition, and Types of Magmas

Laboratory experiments have established the temperatures and pressures at which different kinds of rock melt, and this information gives us some idea of where melting may take place. Mixtures of sedimentary rocks, for example, melt at temperatures several hundred degrees lower than the melting point of basalt. This information leads us to expect that basalt may start to melt near the base of the crust in tectonically active regions of the upper mantle and that sedimentary rocks melt at shallower depths than basalt. The geometry of plate motions is the link we need to tie tectonic activity and rock compositions to melting (Figure 4.8). Two types of plate boundaries are associated with magma formation: mid-ocean ridges, where the divergence of two

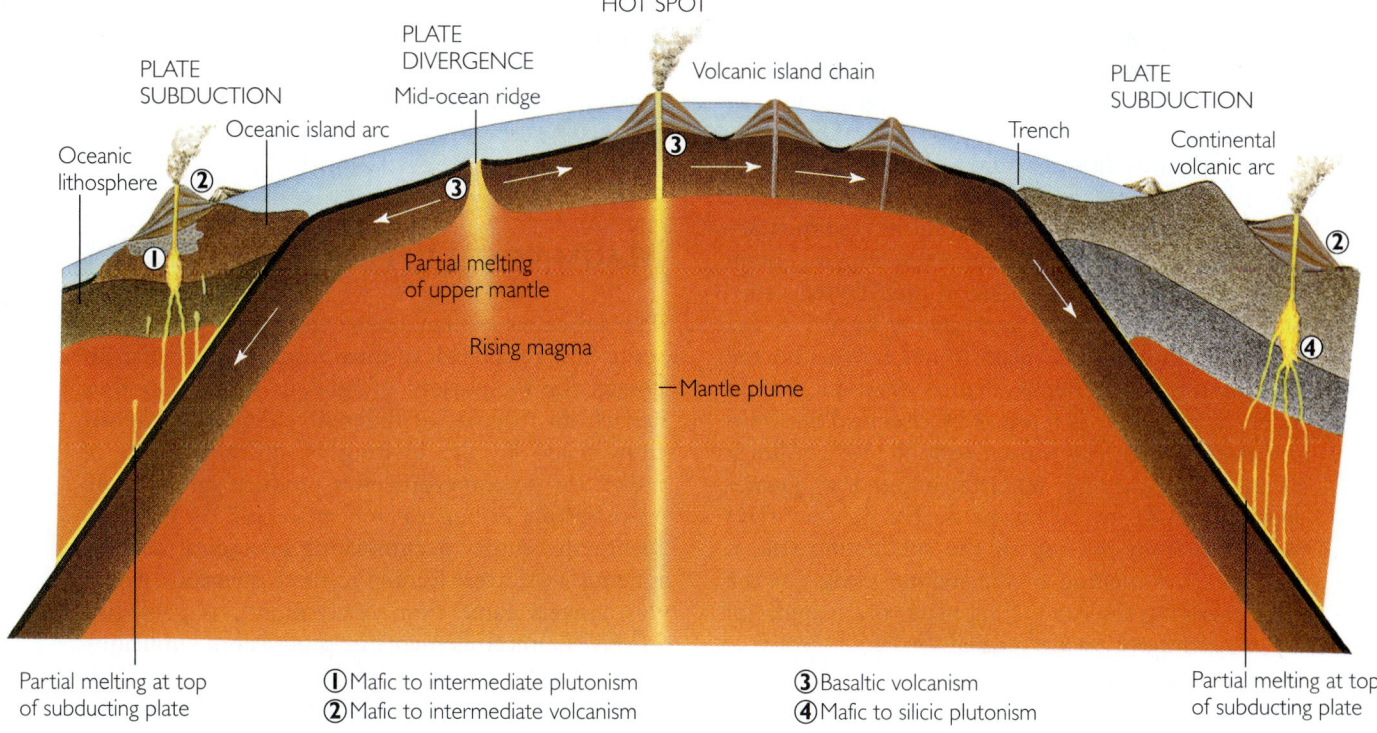

FIGURE 4.8 The three main types of magma—basaltic, andesitic, and rhyolitic—form under conditions that are strongly connected to movements of lithospheric plates. These movements control where rocks of the crust and upper mantle melt and whether they will be intruded or extruded.

plates causes the seafloor to spread, and subduction zones, where one plate dives beneath another.

Before proceeding, recall the terminology used to describe igneous rocks and remember that rocks can have identical compositions but different textures, depending on their intrusive (plutonic) or extrusive (volcanic) form. Geologists label magmas in the same way, using the names of corresponding igneous rock groups. The names most commonly used are those of the volcanic rocks, such as rhyolitic (the felsic group), andesitic (the intermediate group), and basaltic (the mafic group). We will use this terminology not only in this chapter but in Chapter 5 and throughout the text.

MID-OCEAN RIDGES At mid-ocean ridges, heat in the form of rising convection currents in the mantle causes the formation of basaltic magma. Basaltic magma forms in the hot upper mantle below mid-ocean ridges; it then rises to collect in shallow, narrow, wedge-shaped magma chambers near the crest of the ridge. Tremendous quantities of basaltic magma flow intermittently from the rifts and fissures of mid-ocean ridges, giving rise to the abundant lavas of the seafloor.

SUBDUCTION ZONES Other kinds of magmas underlie regions in which volcanoes are highly concentrated, such as the Andes Mountains of South America and the Aleutian Islands of Alaska. Both of these regions were generated by the subduction of one plate under another. The magmas of subduction zones form partly from a mixture of seafloor sediments and partly from basaltic and felsic crust. Sediments have some water remaining in pore spaces. In addition, shales, the most abundant sedimentary rocks, are very high in clay minerals, which contain much water chemically bound in their crystal structure. Sediments become very deeply buried as the subducting lithospheric plate moves down into the lower crust. At moderate depths of about 5 km, much of this water is released by chemical reactions as the temperature increases to about 150°C. Almost all of the remaining water is released at greater depths, from 10 to 20 km or deeper. As this water moves up from the top of the subducting slab, it promotes the melting of the mantle wedge of the plate overlying the subducting plate, and magmas of varying composition are formed.

The compositions of the sedimentary and basaltic and felsic materials that become part of the magma determine the types of igneous rock that can form. The igneous rocks of these subduction zones are generally more silicic than the basalts of mid-ocean ridges. They include much andesite and lesser amounts of more felsic volcanic rocks.

Deep in the crust, beneath the volcanoes, intrusive rocks of intermediate to silicic compositions—from diorite to granite—are formed at the same time that magmas erupt at Earth's surface. These intrusives are added to the base of the crust, thickening it by a process called *underplating*.

MANTLE PLUMES Basalts similar to those produced at mid-ocean ridges are found in thick accumulations over some parts of continents distant from plate boundaries. In the states of Washington, Oregon, and Idaho, the Columbia and Snake rivers flow over a great area covered by this kind of basalt, which solidified from lavas that flowed out millions of years ago. Large quantities of basalt are also erupted in isolated volcanic islands far from plate boundaries, such as the Hawaiian Islands. In such places, slender, pencillike *plumes* of hot basaltic magma rise from deep in the mantle, perhaps as deep as the boundary of the core and the mantle. Mantle plumes, most of them far from plate boundaries, are the "hot spots" of the Earth and are responsible for the outpouring of huge quantities of basalt.

To sum up, basaltic magmas form in the upper mantle beneath mid-ocean ridges and in the lower mantle beneath intraplate hot spots. Magmas of varying composition form at subduction zones, depending on how much felsic material and water the rocks overlying the subduction zone contribute to the melt.

MAGMATIC DIFFERENTIATION

The processes we have been discussing all demonstrate how rocks melt to form magmas. But what accounts for the variety of igneous rocks? Do they arise from magmas of different chemical compositions made by the melting of different kinds of rocks? Or do some processes produce variety from an originally uniform parent material? By the early twentieth century, accumulating geological data on igneous rocks were leading to these questions. Again, the answers came from experiments, as geologists mixed chemical elements in proportions simulating those of natural igneous rocks and melted them in high-temperature furnaces. As the melts subsequently cooled and solidified, geologists carefully observed the temperatures at which crystals formed and they recorded the chemical compositions of those crystals. This research gave rise to the theory of **magmatic differentiation,** a process by which a uniform parent magma may lead to rocks of a variety of compositions. Magmatic differentiation occurs because different minerals crystallize at different temperatures. During the crystallization process the composition of the magma changes as it is depleted of the chemical elements withdrawn to make the crystallized minerals.

In a mirror image of partial melting, the first minerals to crystallize from the cooling molten rock were the last to melt in partial-melting experiments. As this initial crystallization withdrew chemical elements from the melt, the magma's composition changed. As cooling continued, the next minerals to crystallize were those that had melted in the same temperature range in melting experiments. Again the magma's chemical composition changed as different elements were withdrawn. Finally, at the temperature at which the magma solidified completely, the last minerals to crystallize were the ones that melted first when the rock was heated to the beginning of its melting temperature range.

In the course of many experiments, two patterns of crystallization emerged:

- *Continuous and gradual change* In this pattern, illustrated by the plagioclase feldspars, the composition of the successively formed feldspars changed continuously and gradually as crystallization proceeded.

- *Abrupt and discrete change* In the other pattern, characteristic of the mafic minerals such as olivine and pyroxene, the composition of the crystals changed discontinuously during cooling, one mineral abruptly changing to another at a particular temperature.

Because these crystallization patterns are so basic to an understanding of magmatic differentiation, we will examine them in more detail.

The Continuous Reaction Series

When melts of various plagioclase feldspar compositions were cooled, the first crystals to form were always richer in calcium than the melt. Their formation partially depleted the melt of calcium, so that the remaining melt became richer in sodium. As a result, when the melt continued to cool, the next crystals to form were more sodium-rich. At the same time, the calcium-rich crystals formed earlier reacted chemically with the now more sodium-rich melt. In this reaction, calcium ions in the crystal were replaced by sodium ions from the melt so that the calcium-rich crystals formed earlier became richer in sodium. All crystals, both earlier and new, now had the same composition (Figure 4.9). As the process continued, both melt and crystals gradually became richer in sodium and poorer in calcium, until, when crystallization was complete, the final

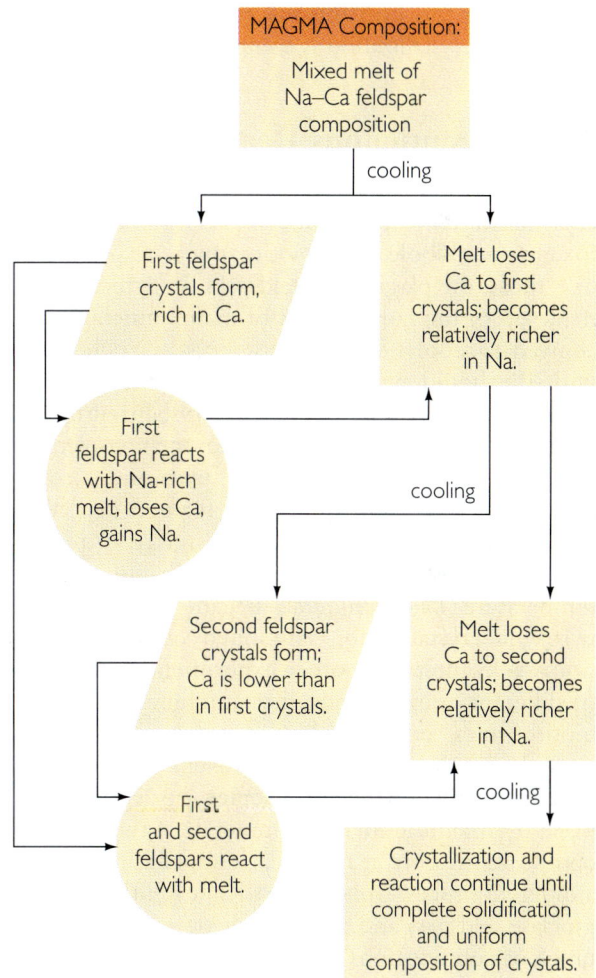

FIGURE 4.9 A continuous reaction series in plagioclase feldspar. As molten plagioclase feldspar cools, crystallization occurs. The crystals that form are always richer in calcium than the liquid they leave behind, which is correspondingly richer in sodium. In successive stages, liquid and crystals interact. Newly formed crystals and earlier crystals react with the liquid, so that at any given time all existing crystals, old and new, have the same composition. At the point of complete solidification, all crystals have reacted to attain a uniform composition identical to that of the original liquid.

homogeneous solid mass of crystals had arrived at the same composition as the original melt. At all times the mineral being crystallized was a plagioclase feldspar.

The key to this process is the continuous reaction of crystals with the melt. Each continuously changes by small amounts, so that at any point in the course of crystallization all crystals have the same composition. Crystals and melt move through a series of compositions, which in the earlier stages are richer in calcium and in later stages are richer in sodium. With continued cooling this **continuous reaction series** proceeds until crystallization is complete.

The Discontinuous Reaction Series

A somewhat different process is involved in the crystallization of mafic minerals, such as olivine, pyroxene, amphibole, and biotite mica. Experiments like those on plagioclase feldspars, in which melts made up of the components of mafic minerals were allowed to cool slowly so that crystals could react with the liquid, also showed a systematic order of crystallization. At 1800°C, olivine crystallized first; it continued to crystallize until the melt cooled to 1557°C. Below that temperature, pyroxene, a completely different mineral, abruptly formed, and all of the earlier olivine crystals were converted to pyroxene (Figure 4.10). At 1543°C, cristobalite, a high-temperature silica mineral, began to form, and pyroxene crystallization continued until solidification was complete. In other experiments with melts of different compositions, first amphibole and then biotite mica crystallized at temperatures successively lower than the olivine-pyroxene series. In this **discontinuous reaction series,** reactions take place between the melt and minerals of two definite compositions only at particular temperatures. This is a different process from the gradual evolution of plagioclase feldspars and parent melt over a continuous range of compositions and temperatures.

The crystal structures of the minerals of the two reaction series are part of the differences in crystallization patterns (see Table 4.1). Throughout the changes in the continuous reaction series, the basic feldspar crystal structure remains constant, although the proportions of calcium and sodium change. In contrast, the crystal structures of the discontinuous reaction series change as temperatures fall, forming increasingly complex arrangements of silica tetrahedra. At the highest temperatures, the crystal structure of olivine is that of isolated silica tetrahedra, the basic building blocks of silicate minerals (see Chapter 2).

In the next stage, the pyroxenes are single chains of tetrahedra. Then come the amphiboles, double chains of tetrahedra, followed by the micas, sheets of tetrahedra. At the end stages of both the continuous and discontinuous series, quartz and the feldspars are three-dimensional frameworks of silica tetrahedra.

In the cooling of a natural magma, which normally contains the chemical elements of both plagioclase feldspars and mafic minerals, both patterns of crystallization go on simultaneously. As the temperature of such a magma drops below 1550°C, for example, pyroxene crystals form through discontinuous reaction and a pure calcium plagioclase feldspar crystallizes through continuous reaction.

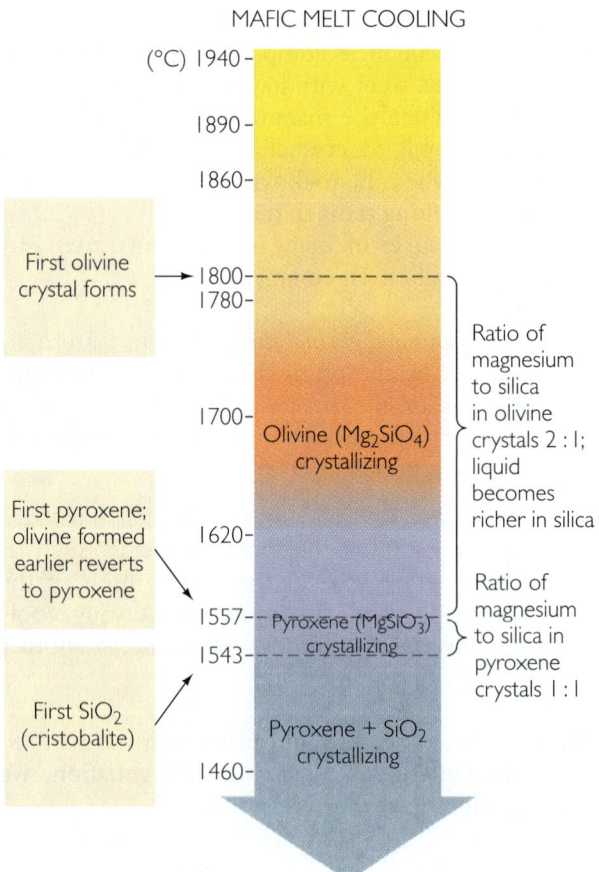

FIGURE 4.10 Part of a discontinuous reaction series. The sequence illustrated here occurs during the crystallization of a cooling liquid of magnesium and silica, in which silica is about 50 percent by weight. This is one part of a discontinuous reaction series by which different minerals successively crystallize from a melt in a series of abrupt changes. Olivine does not first crystallize at exactly 1800° in every melt; the exact temperature depends on the composition of the cooling melt.

Although these two reaction series explain the composition of many igneous rocks, they fall short of accounting for many others. Consider a natural magma in which all crystals have reacted completely with the liquid rock at all stages of crystallization. Given these conditions, we would expect to find only the final products of a crystallization—a single plagioclase feldspar corresponding to the composition of the original melt and a pyroxene. No traces of the first crystallization products—the original calcium-rich feldspar and the original olivine—would be left. Geologists examining volcanic rocks knew some part of the theory was missing when they found many such rocks with calcium-rich plagioclases and with olivine.

Fractional Crystallization

The theory of magmatic differentiation needed one more essential piece: a way to account for the preservation of minerals formed earlier as the composition of the melt changed. N. L. Bowen, a Canadian geologist, proposed such a mechanism about 75 years ago. Even as an undergraduate, Bowen had been interested in the chemical basis for igneous rock formation. Later he focused on the continuous and discontinuous crystallization series. He was especially interested in the course of crystallization in situations in which plagioclase feldspars—or mafic minerals—failed to change composition through reaction with the remaining liquid. Such might be the case, for example, if a magma cooled more rapidly than usual. In such a magma, plagioclase feldspar crystals would have time to grow, but only the outer surfaces of existing crystals would have enough time to react with the changing liquid. As a result, only the outer layer of each crystal would change composition. As crystallization proceeds, each successive layer of feldspar is covered by plagioclase that is progressively richer in sodium.

The end product would be what we now call a **zoned crystal,** a single crystal of one mineral that has a different chemical composition in its inner and outer parts (Figure 4.11). In Bowen's example, the crystals' compositions would change in a series of gradual steps from calcium-rich interiors to sodium-rich exteriors. There is more to the problem of limited reaction than the effects of rapid crystallization, however. If the calcium-rich centers of the growing crystals were encapsulated, the liquid would not reach a state of equilibrium with crystals, as it would during a slow continuous reaction. The liquid would remain rich in sodium because the calcium from the

FIGURE 4.11 A zoned crystal. The colors in this crystal correspond to differences in chemical composition in various zones within the crystal. Each zone represents a fractional crystallization stage, during which the earlier-formed inner zones did not react with the liquid. *(Chip Clark.)*

crystal interiors would not be available to replace the sodium in the melt.

Relying on both experimentation and field observation, Bowen proposed a new theory of magmatic differentiation. Though the mechanisms he suggested are no longer accepted for the differentiation of most igneous rocks, his ideas served as a foundation for most later work and still teach us much today. Bowen proposed a process by which the first-formed crystals would be segregated from the remaining melt. This segregation could happen in several ways. For example, crystal settling might cause early crystals formed in a magma chamber to settle to the chamber's floor and thus be removed from further reaction with the remaining liquid. Another possibility is that structural deformation during the crystallization process might squeeze the remaining liquid from the chamber, segregating and compressing the crystals as a distinct intrusive body. The magma would then migrate to new locations, forming new chambers.

By either scenario, settling of crystals or structural deformation, crystals that had formed early would be segregated from the remaining melt, which would then behave as though it had just begun to crystallize. In the continuous reaction series, for example, the melt, already enriched in sodium at that point, would start to crystallize a feldspar much richer in sodium than any that would have crystallized from an unsegregated magma. Continued crystallization would produce a mass of such feldspars

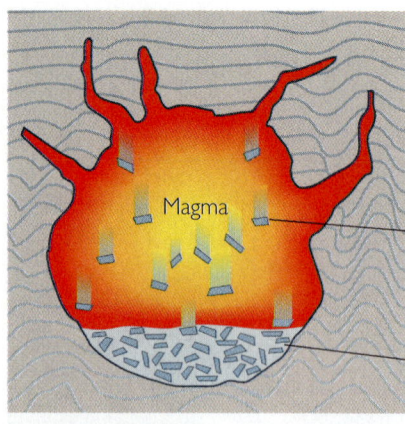

(a) Early crystallization

Magma

Crystals form from magma cooling and settle to floor of chamber

Crystals from early cooling accumulate

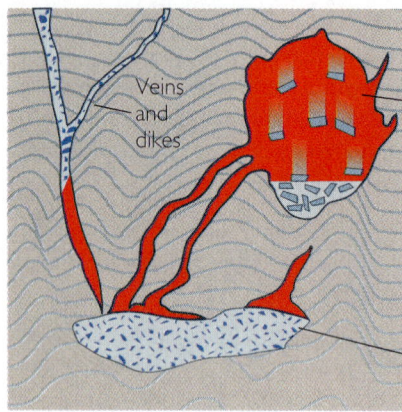

(b) Later deformation squeezes remaining liquid from crystal mush

Veins and dikes

Magma migrates to secondary chamber, where it continues to crystallize

Mass of crystals formed early are segregated and compressed to form separate intrusive body

FIGURE 4.12 Two stages in the evolution of a magma differentiated by fractional crystallization. In the first stage (a), early crystals settle to the floor of the magma chamber. As cooling proceeds, structural deformation may squeeze the remaining liquid from the chamber, segregating and compressing those crystals to form a separate intrusive body (b). The separated liquid migrates to form veins, dikes, and other magma chambers in new locations, where it continues to crystallize.

that would be much richer in sodium than the rock that the original melt would have produced. Meanwhile, the segregated calcium-rich crystals that had formed first would form a mass of feldspar much richer in calcium than the original melt. **Fractional crystallization** is the term used for this separation and removal of successive fractions of crystals formed from a cooling magma (Figure 4.12). Bowen believed this process could account for the preservation of early-formed calcium-rich feldspars and the crystallization of sodium-rich plagioclases from an originally calcium-rich magma.

Bowen went on to show that fractional crystallization could work with the discontinuous mafic mineral series as well. Just as the first plagioclase feldspars to crystallize may be removed, the first-formed crystals of olivine in a discontinuous reaction series may settle out and so be removed from further reaction. These mafic minerals would be found with their corresponding plagioclase feldspars. With the first-formed olivine removed, the magma would then crystallize pyroxene. Thus both continuous and discontinuous crystallization paths might produce a range of products similar to those found in natural igneous rocks. But like any scientific theory, the theory of fractional crystallization had to be tested before it could be accepted.

From Laboratory to Field Observation: The Palisades Intrusion

A perfect test case for the theory of fractional crystallization was the Palisades, a massive cliff facing the city of New York on the west bank of the Hudson River. The Palisades is an igneous formation about 80 km long and in places more than 300 m from top to bottom. It was intruded as a melt of basaltic composition into almost horizontal sedimentary rocks, and it contains abundant olivine near the bottom, pyroxene and plagioclase feldspar in the middle, and mostly plagioclase feldspar near the top (Figure 4.13a). The variation in mineral composition from top to bottom made this formation a perfect site for testing Bowen's theory and showed how laboratory experiments could help explain field observations.

From experiments on the melting of rocks with about the same proportions of the various minerals found in the Palisades intrusion, geologists knew that the temperature of the melt had to have been about 1200°C. The parts of the magma within a few meters of the relatively cold upper and lower contacts of the surrounding sedimentary rocks cooled quickly, forming a fine-grained basalt and preserving the chemical composition of the original melt. But the hot interior of the intrusion cooled more slowly, as the slightly larger crystals testify.

Bowen's experiments on fractional crystallization lead us to think that the first mineral to crystallize from the slowly cooling interior would be olivine, a heavy mineral that would sink through the melt to the bottom of the intrusion. There it can be found today in a coarse-grained, olivine-rich layer just above the chilled, fine-grained basaltic layer along the bottom contact in the Palisades (Figure 4.13b). Continued cooling would produce pyroxene

(a)

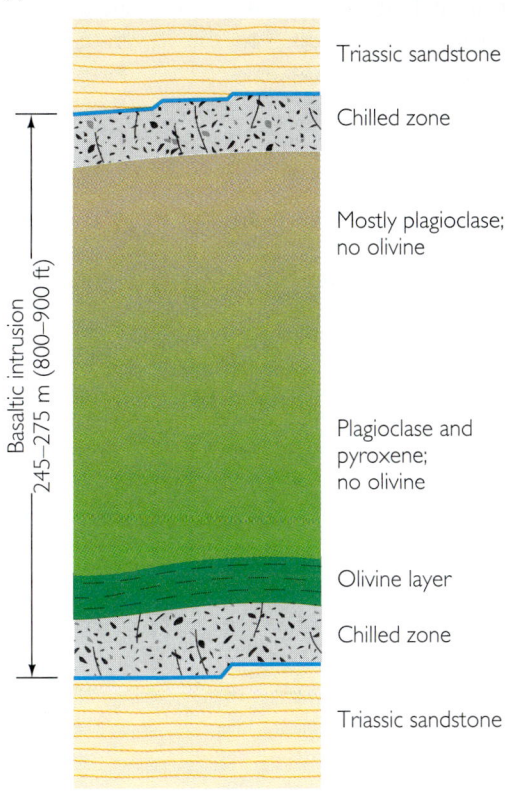

(b)

FIGURE 4.13 The Palisades demonstrates fractional crystallization. (a) The Palisades is the result of an intrusion of basaltic magma into sedimentary rocks some 200 million years ago along what is now the Hudson River. *(Breck P. Kent.)* (b) In a classic application of laboratory science to field observation, geologists used the concept of fractional crystallization to interpret the Palisades' vertical variation in texture and mineral composition.

position until successive layers of settled crystals would be topped off by a layer of mostly plagioclase, sodium-rich feldspar crystals.

The explanation of the layering of the Palisades intrusion as the result of fractional crystallization was an early success of the first version of the theory of magmatic differentiation. It firmly tied field observations to laboratory experiments and was solidly based on chemical knowledge. More than two-thirds of a century of geological research has passed since the Palisades was first seen as a test case, and we now know that the Palisades has a more complex history, including several injections of magma and a more complicated story of olivine settling. Nevertheless, the general picture remains a valid instance of fractional crystallization.

crystals, followed almost immediately by calcium-rich plagioclase feldspar. These minerals, too, had settled out through the magma to accumulate in the lower third of the Palisades intrusion. The abundance of plagioclase feldspar in the upper parts of the intrusion corresponded with the expectation that the melt would then continue to change com-

Bowen's Theory of Magmatic Differentiation

If fractional crystallization and magmatic differentiation were to account for the differences among igneous rocks, they would have to explain two seemingly contradictory facts:

- The widespread distribution and abundance of granite. This intrusive rock is at the silicic end of the igneous rocks, and it contains abundant sodium-rich plagioclases and other minerals with low melting temperatures.

- The equally widespread distribution and abundance of basalt. Basalt, an extrusive rock, is at the mafic, or low silica, end of the spectrum; it contains calcium-rich plagioclase feldspars and other minerals with high melting temperatures.

Studies of the lavas of volcanoes showed that basaltic magmas are common, far more common than the rhyolitic magmas that correspond in composition to granites. How could the abundant granites have been derived from basaltic magmas?

Bowen's idea was that an originally basaltic magma would gradually cool and differentiate to a more silicic, lower-temperature melt by fractional crystallization. Early stages of differentiation of a basaltic magma by fractional crystallization would produce andesitic magma, which might erupt to form andesite lavas or solidify by slow crystallization to form diorite intrusives. Intermediate stages would make magmas of granodiorite composition. If this process were carried far enough, its late stages would form rhyolite lavas and granite intrusions.

Bowen's Reaction Series

In 1928 Bowen capped more than 10 years of active experimentation by proposing a simplified general scheme for magmatic differentiation that combined the continuous and discontinuous fractional crystallizations of the major minerals of igneous rocks. The **Bowen reaction series,** as it is called, starts with the cooling of a high-temperature basaltic magma that gradually differentiates by fractional crystallization along two paths simultaneously (Figure 4.14):

1. The continuous path of the plagioclase feldspars, starting at high temperatures with calcium-rich feldspar and proceeding to the lower-temperature sodium-rich feldspar.

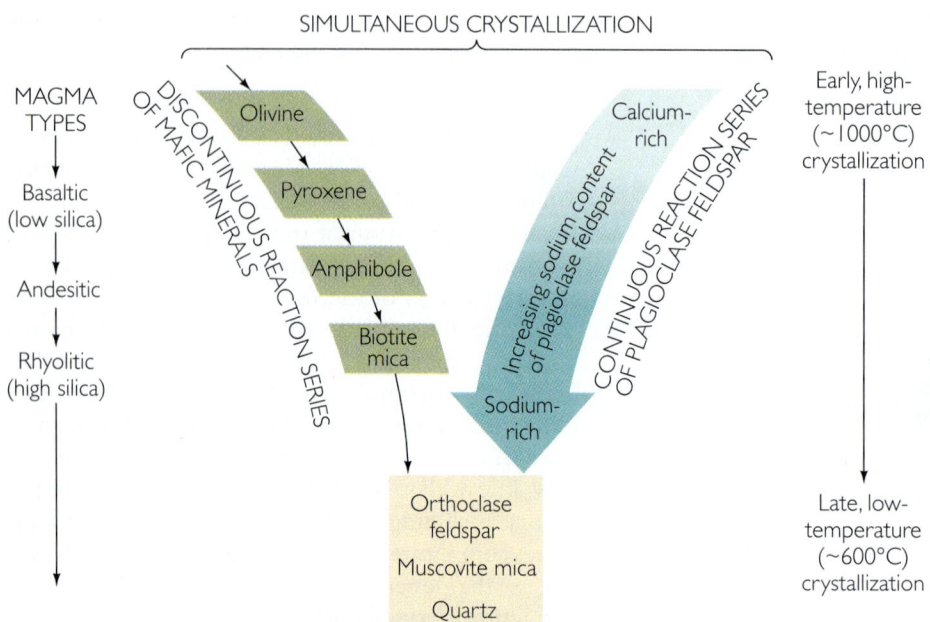

FIGURE 4.14 Bowen's reaction series. Bowen proposed this simplified general scheme to show how the sequence of fractional crystallization of a melt could lead to the formation of differentiated magmas. Two different but simultaneous paths would allow a variety of minerals to crystallize during the cooling of a high-temperature basaltic magma. Though this series does describe what happens to a hypothetical magma, it does not adequately explain many examples of igneous rock origin, such as the widespread intrusion of granite. Nevertheless, it remains a valuable chart for understanding the role of fractional crystallization.

2. The discontinuous path of the mafic minerals, starting at the high-temperature end with olivine, then progressing to pyroxene, amphibole, and biotite mica as the magma cools.

The paths of the two series converge at a final, low-temperature (about 600°C) magma crystallizing the minerals of granite: albite (sodium-rich plagioclase feldspar) and orthoclase (potassium feldspar), muscovite mica, and quartz.

To sum up the story thus far, a train of observations and experiments led to the Bowen reaction series as a possible explanation for the great variety of igneous rocks:

- Melting and crystallization behavior of rocks and melts revealed the continuous and discontinuous reaction series.

- Fractional crystallization explained how magmas could evolve through a series of compositions by the separation of the crystals formed earlier from the magma.

- The continuous and discontinuous series were put together as an explanation for the evolution of granites and other intrusive and extrusive rocks from an originally basaltic magma.

Modern Theories since Bowen

At first Bowen's theory of magmatic differentiation seemed to be a great success. It explained well how different types of igneous rocks could form by fractional crystallization, and it provided an understanding of the kinds of rocks seen in the field, such as the Palisades mafic intrusion. Bowen's theory also seemed to explain how rhyolite, the extrusive equivalent of granite, could form toward the end of a series of eruptions that started with basaltic lavas.

As often happens when a scientific theory quickly dominates an area of science, further work showed the need to modify and add to Bowen's original theory of magmatic differentiation. One line of research showed that such great lengths of time would be needed for small crystals of olivine to settle through a dense, viscous magma that they might never reach the bottom of a magma chamber. Other researchers demonstrated that many layered intrusions similar to but much larger than the Palisades do not show the simple progression of layers predicted by Bowen's theory.

The biggest problem, however, was the source of granite. The first sticking point is that the great volume of granite found on Earth could not have been formed as Bowen's reaction series suggests. Large quantities of liquid volume are lost by crystallization during successive stages of differentiation. To produce the existing amount of granite, an initial volume of basaltic magma 10 times the size of a granitic intrusion would be required. That abundance would imply the crystallization of huge quantities of basalt underlying granite intrusions. But geologists could not find anything like that amount of basalt. Even where great volumes of basalts are found—at mid-ocean ridges—there is no wholesale conversion to granite through magmatic differentiation.

Most in question is Bowen's original idea that all granitic rocks evolve from the differentiation of a single type of magma, a basaltic melt. Instead, geologists discovered that melting of varied source rocks of the upper mantle and crust is responsible for much variation in magma composition, as follows:

1. Rocks in the upper mantle might partially melt to produce basaltic magma.

2. A mixture of sedimentary rocks and basaltic oceanic rocks such as those found in subduction zones might melt to form andesitic magma.

3. A melt of sedimentary, igneous, and metamorphic continental crustal rocks might produce granitic magma.

Magmatic differentiation does operate, but its mechanisms are more complex than Bowen recognized. Here are just a few of the points that have altered Bowen's original theory:

- Partial melting is of great importance in the production of magmas of varying composition. Magmatic differentiation can be achieved by partial melting of mantle and crustal rocks over a range of temperatures and water contents. Thus basaltic magma can be formed by a 10 to 15 percent partial melting of rocks of the upper mantle at depths of around 100 km. An andesitic magma can be formed by partial melting of a water-rich basaltic oceanic crust that heats up as it descends along a subduction zone. A rhyolitic magma may be formed by partial melting in the lower crust of a mixture of continental crustal rocks or, alternatively, of andesite. In all of these cases one can apply

Bowen's reaction series in reverse to predict the composition of the magma as it is formed from partial melting.

- Magmas do not cool uniformly; they may exist at a wide range of temperatures within a magma chamber.

- The differences in temperature in magma chambers may cause the chemical composition of the magma to vary from one region to another.

- A few melt compositions are immiscible—they do not mix with one another, just as oil and water do not mix. When such magmas coexist in one magma chamber, each forms its own crystallization products.

- Some magmas that are miscible may give rise to a crystallization path different from that followed by any one magma alone.

We also now know more about the physical processes that interact with crystallization within magma chambers (Figure 4.15). Magma at various temperatures in different parts of a magma chamber may flow turbulently, crystallizing as it circulates. Crystals may settle, then again be caught up in currents, and eventually be deposited on the chamber's walls. The margins of such a magma chamber may be a "mushy" zone of crystals and melt lying between the solid rock border of the chamber and the completely liquid magma within the main part of the chamber. And at some mid-ocean ridges, such as the East Pacific Rise, a mushroom-shaped magma chamber may be surrounded by hot basaltic rock with only small amounts (1 to 3 percent) of partial melt.

Bowen's original theory of magmatic differentiation has been supplanted since he proposed it many decades ago. Nevertheless, as we noted earlier, most later work on the differentiation of igneous rocks was built on the foundation of Bowen's ideas.

FORMS OF MAGMATIC INTRUSIONS

As we noted earlier, geologists cannot directly observe the forms that intrusive igneous rocks take when magmas intrude the crust. We can only deduce their shapes and distributions from evidence gained in geological fieldwork done millions of years after the rocks were formed, long after the magma cooled and the rocks were uplifted and exposed to erosion.

To be sure, we do have indirect evidence of current magmatic activity. Earthquake waves, for example, show us the general outlines of magma chambers that underlie some active volcanoes, but they cannot tell us the shape or size of intrusions supplied from those magma chambers. In some nonvolcanic but tectonically active regions, such as an area near the Salton Sea in southern California, high temperatures in deep drill holes reveal a crust much hotter than normal, which may be evidence of an intrusion at depth.

But in the end, most of what we know about intrusive igneous rock is based on the work of field geologists who have mapped and compared a wide variety of outcrops and have reconstructed their history. Their studies have resulted in the description and classification of the many irregular and variable forms of intrusive bodies. In the following pages, we consider some of these bodies—plutons, sills and dikes, and veins. Figure 4.16 illustrates a variety of extrusive and intrusive structures.

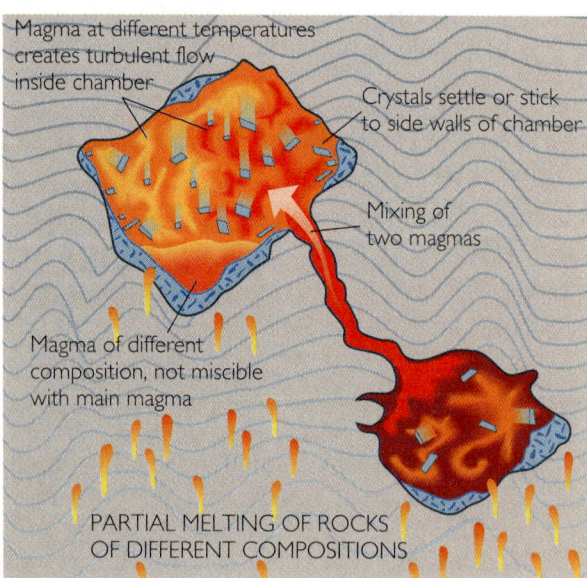

FIGURE 4.15 Modern ideas of magmatic differentiation. Geologists now recognize that magmatic differentiation does operate, but its mechanisms are more complex than Bowen recognized. Melting is usually partial. Some magmas derived from rocks of varying compositions may mix, while other magmas are immiscible. Crystals may be transported to various parts of the magma chamber by turbulent currents in the liquid.

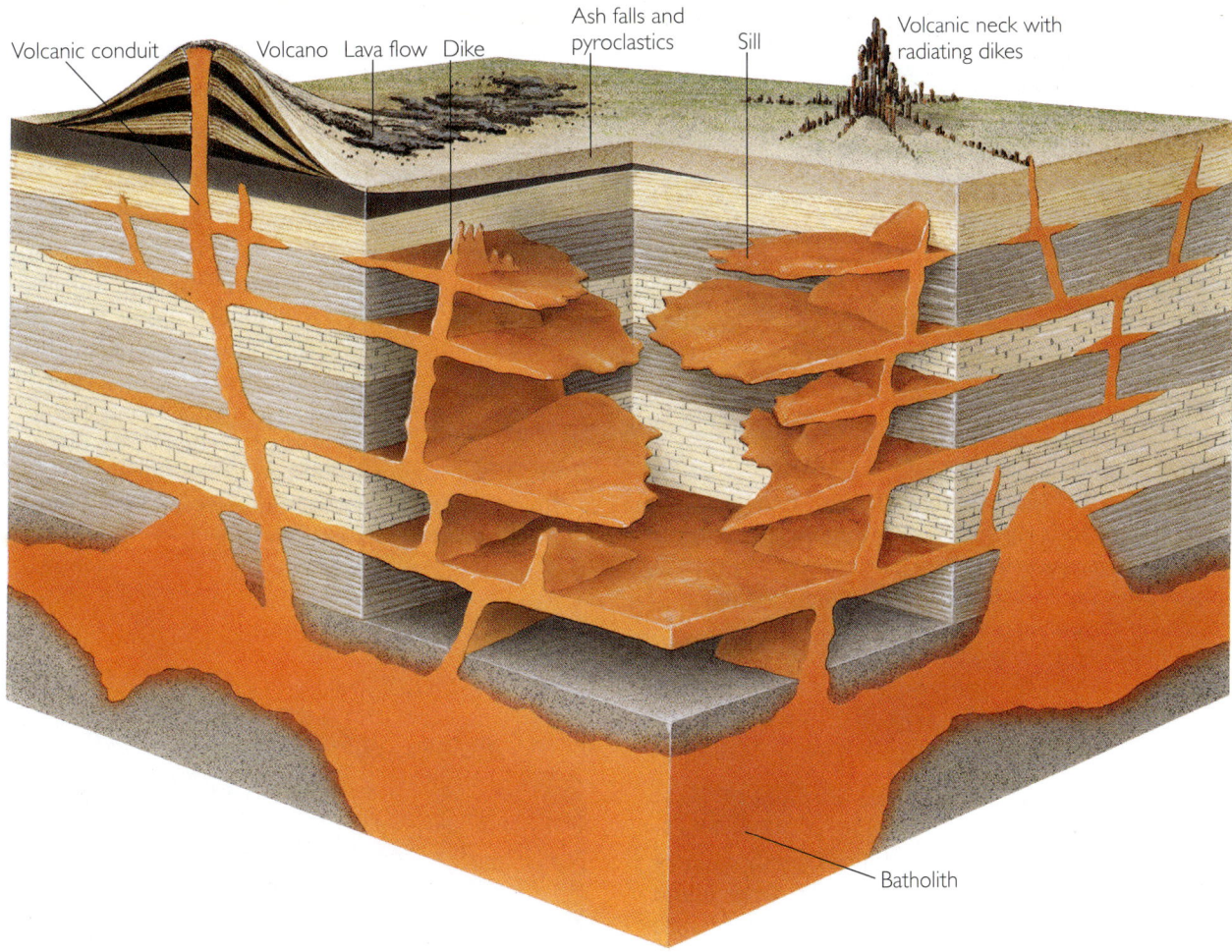

FIGURE 4.16 Basic extrusive and intrusive igneous structures. Notice that dikes cut across layers of country rock, but sills run parallel to them. Batholiths are the largest forms of plutons.

Plutons

Plutons are large igneous bodies that formed at depth in Earth's crust; they range in size from 1 km^3 to hundreds of cubic kilometers. These large bodies become accessible to study when uplift and erosion uncover them or when mines or drill holes cut into them. Plutons are highly variable, not only in size but also in shape and in their relation to the surrounding country rock.

This wide variability in part reflects the different ways magma makes space for itself as it rises through the crust. Most magmas intrude at great depths—deeper than 8 to 10 km. At these depths, few holes or openings exist because the great pressure of the overlying rock would close them. But even that high pressure is overcome by the presence of the upwelling magma.

Magma rising through the crust makes space for itself in three ways:

- *Wedging open the overlying rock* As the magma lifts that great weight, it fractures the rock, penetrates the cracks, wedges them open, and so flows into the rock. Overlying rocks may bow up during this process.

- *Breaking off large blocks of rock* Magma can push its way upward by breaking off blocks of the invaded crust. These blocks sink into the magma, melt, and blend into the liquid, in some places changing the composition of the magma.

- *Melting surrounding rock* Magma also makes its way along by melting walls of country rock.

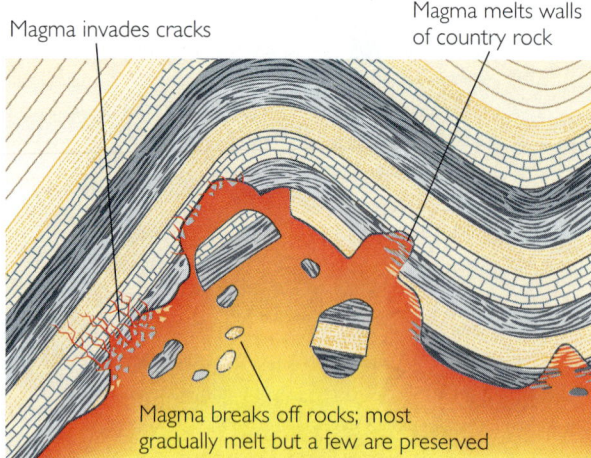

FIGURE 4.17 Magmas make their way into country rock in three basic ways: by invading cracks and wedging open overlying rock, by breaking off rock, and by melting surrounding rock. Here we see a magma intruding an area of surrounding folded rock.

Figure 4.17 illustrates these three methods of magmatic intrusion.

Most plutons show sharp contacts with country rock and other evidence of intrusion of a liquid magma into solid rock. Other plutons grade into country rock and show structures vaguely resembling those of sedimentary rocks. This latter set of features suggests that these plutons formed from preexisting sedimentary rocks that underwent granitization. **Granitization** is the process by which granite is formed from other rocks by recrystallization, with or without complete melting. (We discuss granitization in more detail in Chapter 8, "Metamorphic Rocks.") This type of pluton was converted to granite by partial melting and invasions by hot solutions and gases percolating up from great depths with the magma.

Batholiths, the largest plutons, are great irregular masses of coarse-grained igneous rock that by definition cover at least 100 km^2 (see Figure 4.16). Similar but smaller plutons are called **stocks.** Both batholiths and stocks are **discordant intrusions;** that is, they cut across the layers of the country rock they intrude.

Batholiths are found in the cores of tectonically deformed mountain belts. Geological field evidence is accumulating to show that batholiths are horizontal sheetlike or lobate thick bodies extending from a funnel-shaped central region. Their bottoms may extend 10 to 15 km deep, and a few are estimated to go even deeper. Batholiths' coarse grain is a consequence of slow cooling at great depths.

Sills and Dikes

Sills and dikes are similar to plutons in many ways, but they are smaller and they have a different relation to the layering of the surrounding intruded rock. A **sill** is a tabular sheetlike body that was formed by injection of magma between parallel layers of preexisting bedded rock (see Figure 4.16). Sills are **concordant intrusions**—that is, their boundaries lie parallel to these layers, whether or not the layers are horizontal. Sills range in thickness from a single centimeter to hundreds of meters, and they can extend over considerable areas. Figure 4.18

FIGURE 4.18 Dark sill formed by magma intruded between layers of sedimentary rock (light-colored bedded rock at top) in Big Bend National Park, Texas. The sill shows columnar jointing. *(Tom Bean.)*

shows a large sill at Big Bend National Park in Texas. The 300-meter-thick Palisades intrusion (see Figure 4.13) is another large sill.

Sills may superficially resemble layers of lava flows and pyroclastic material, but they differ from these layers in four ways:

- They lack the ropy, blocky, and vesicle-filled structures that characterize many volcanic rocks.

- They are more coarsely grained than volcanics because the sills have cooled more slowly.

- Rocks above and below sills show the effects of heating; their color may have been changed, or they may have been mineralogically altered by contact metamorphism.

- Many lava flows overlie weathered older flows or soils formed between successive flows; sills do not.

Dikes are the major route of magma transport in the crust. They are like sills in being tabular igneous bodies, but dikes cut across layers of bedding in country rock, rather than run parallel to them (see Figure 4.16). Dikes sometimes form by forcing open fractures that formed earlier, but they more often create channels through new cracks opened by the pressure of magmatic injection. Some individual dikes can be followed across country for tens of kilometers. Their widths vary from many meters to a few centimeters. In some dikes, one can see *xenoliths*—fragments of country rock completely surrounded by the intrusive material. These fragments, which once floated in the intruding magma, are good evidence of disruption of the surrounding rock during the intrusion process (Figure 4.19). Dikes rarely occur alone; more typically large numbers, or swarms, of hundreds or thousands of dikes are found in a region that has been deformed by a large igneous intrusion.

The texture of dikes and sills varies. Many are coarsely crystalline, with an appearance typical of

FIGURE 4.19 Dike of igneous rock (dark) intruded into shaly sedimentary rock (reddish brown) in Grand Canyon National Park, Arizona. *(Tom Bean/DRK.)*

intrusive rocks. Many others are finely grained and look much more like volcanic rocks. Because texture reflects rate of cooling, we know that the fine-grained dikes and sills invaded country rock nearer the Earth's surface, where rocks are cold in comparison with intrusions. Their fine texture is a result of fast cooling. The coarse-grained ones formed at depths of many kilometers and invaded warmer rocks whose temperatures were much closer to their own.

Veins

Veins are deposits of minerals found within a rock fracture that are foreign to the host rock. Irregular pencil- or sheet-shaped veins branch off from the tops and sides of many intrusive bodies. They may be a few millimeters to several meters across, and they tend to be tens of meters to kilometers long or wide. The famous Mother Lode of the Gold Rush of 1849 in California is a vein of quartz bearing crystals of gold. Veins of extremely coarse-grained granite cutting across a much finer-grained country rock are called **pegmatites** (Figure 4.20). They are crystallized from a water-rich magma in late stages of solidification. Pegmatites provide ores of many rare elements, such as lithium and beryllium.

Some veins are filled with minerals that contain large amounts of chemically bound water and that are known to crystallize from hot-water solutions. From laboratory experiments, we know that these minerals crystallize at high temperatures—typically from 250 to 350°C—but not nearly as high as the temperatures of magmas. The solubility and composition of the minerals in these **hydrothermal** (from Greek *hydōr*, water, and *thermē,* heat) **veins** indicate that abundant water was present as the veins formed. Some of the water may have come from the magma itself, but some may be from underground water in the cracks and pore spaces of the intruded rocks. Groundwaters originate as rainwater seeps into the soil and surface rocks. Hydrothermal veins are abundant along mid-ocean ridges. In these areas, seawater infiltrates cracks in basalt and circulates down into hotter regions of the basalt ridge, emerging at hot vents on the seafloor in the rift valley between the spreading plates. (The geology of hydrothermal veins and the valuable ores they contain are discussed in detail in Chapter 23.)

IGNEOUS ACTIVITY AND PLATE TECTONICS

Since the advent of the theory of plate tectonics in the 1960s, geologists have been trying to fit the facts and theories of igneous rock formation into its framework (see Figure 4.8). We noted that batholiths, for example, are found in the cores of many mountain ranges. These mountain ranges were formed by the convergence of two plates. This observation implies a connection between plutonism and the mountain-making process and between both of them and the forces responsible for plate movements—plate tectonics.

The major sites of igneous rock formation are divergence zones—mid-ocean ridges. At these locations, basaltic magmas derived from partial melting of the mantle wells up along rising convection currents. Magma is extruded as lavas, fed by magma chambers below the ridge axis. At the same time, gabbroic intrusions are emplaced at depth (Figure 4.21).

Subduction zones, where one plate dives below another, are major sites of rock melting. The top of a subducting lithospheric plate includes oceanic crust, which is largely basalt formed originally at a mid-ocean ridge. The plate also carries water and still-soft oceanic sediment, which it accumulated during its travels from mid-ocean ridge to subduction zone. As the plate moves downward, it encounters increasing temperature and pressure, which convert the sediments first to sedimentary rocks and then at greater

FIGURE 4.20 A granite pegmatite dike. The center of the dike displays the coarse crystallinity associated with slow cooling. The finer crystals along the boundaries of the dike cooled more rapidly. *(Martin Miller.)*

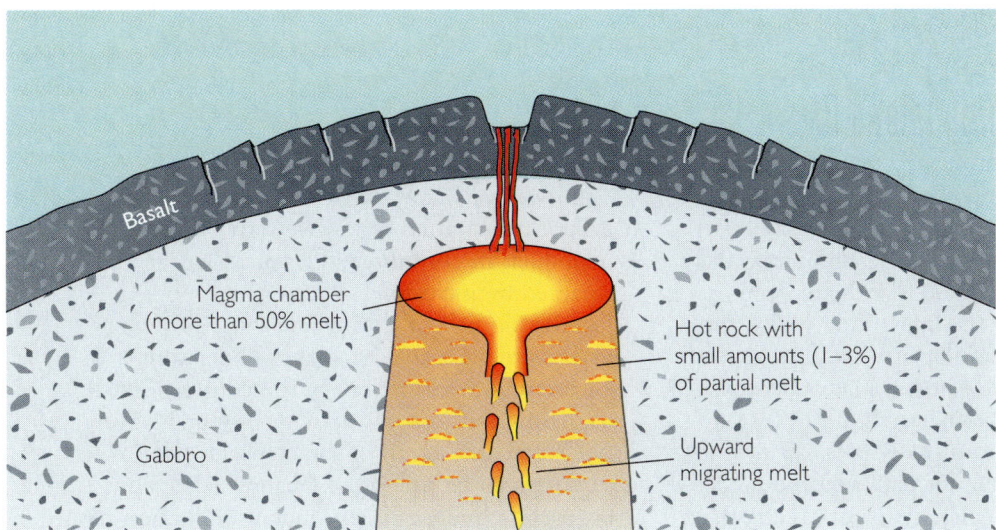

FIGURE 4.21 This diagrammatic cross section of a mid-ocean ridge shows a mushroom-shaped magma chamber below the ridge's central axis. Surrounding the chamber is a thick zone of hot rock containing small amounts of partial melt.

depths to metamorphic rocks. Because of the relatively large amounts of water they contain, these materials have lower melting temperatures than dry crustal or mantle rock. As the lithospheric plate moves deeper, temperatures rise, reaching the melting point of the sedimentary or metamorphic rocks. Continuing downward, the plate finally encounters temperatures that melt the top parts of the basalt. Subduction thus creates magma, or perhaps magmas of several kinds.

As the magmas and water from dehydration reactions work their way upward from the top of the melting subducting slab, they may melt portions of the overriding plate and change their composition. At the same time, the magmas may differentiate by fractional crystallization. The result is a range of igneous rocks, both intrusives and extrusives. Volcanoes over the deeper parts of the subduction zone, where melting is going on, extrude basaltic, andesitic, and rhyolitic lavas and pyroclasts, forming a wide range of volcanic rocks. These volcanoes and the volcanics they expel form the islands of oceanic volcanic arcs, such as the Aleutian Islands of Alaska.

Where subduction takes place beneath a continent, the many volcanoes and the volcanic rocks coalesce to form a mountainous arc on land. Subduction of an offshore oceanic plate has generated one such arc—the Cascade Range with its active volcanoes, such as Mount St. Helens—in northern California, Oregon, and Washington.

As mountains are forming above, intrusive magma bodies are crystallizing deep below, forming igneous rocks that vary from mafic to felsic, depending on the magma's composition and degree of differentiation. Working backward from the patterns and compositions of igneous rocks, we can estimate the composition of the parent magma and the depth of the descending slab. In so doing, geologists can reconstruct events that occurred millions of years ago in the subduction zone (see Feature 4.1).

The terrain of the Japanese Islands is a prime example of the complex of intrusives and extrusives that evolves over many millions of years at a subduction zone. Everywhere in this small country are all kinds of extrusive igneous rocks of various ages, intercalated with mafic and intermediate intrusives, metamorphosed volcanic rocks, and sedimentary rocks derived from erosion of the igneous rocks. The erosion of these kinds of rocks has contributed to the distinctive landscapes portrayed in so many classical and modern Japanese paintings.

In all of these ways, igneous rocks reflect the major forces shaping the Earth. Each plate-tectonic setting produces its own pattern of igneous rocks: the lava flows and pyroclastics extruded from volcanoes; the batholiths, dikes, and sills intruded at depth; and the wide variety of rocks that come from magmas of distinctive compositions following their own routes of differentiation.

4.1 INTERPRETING THE EARTH

Japan, a Growing Island Arc

Japan lies at the intersection of three converging oceanic plates, the giant Pacific and Eurasian plates and the small Philippine Plate. Just east of the Japanese Islands are deep trenches marking the lines along which the Pacific and Philippine plates subduct beneath the Eurasian Plate. As these plates slip downward, they provoke the earthquakes that are so prevalent throughout the islands. Japan is dotted with active and dormant volcanoes, the most famous of which, Fujiyama, is a traditional object of reverence.

Japanese geologists have worked out the history of this subduction complex by mapping the intrusive and extrusive rocks of various ages to show that originally the archipelago was a narrow arc of small islands more like the Marianas of the western Pacific. As subduction continued, the volcanism, accompanied by intrusions at depth, built up a widening belt of land, while some of the igneous rocks emplaced earlier underwent structural deformation. In the course of this igneous and tectonic activity, mountains were elevated, one chain of them spectacular enough to be called the Japanese Alps. Thus the consequence of continued subduction has been the growth of sizable islands that through magmatic differentiation, structural deformation, and sedimentation have come to resemble tiny continents.

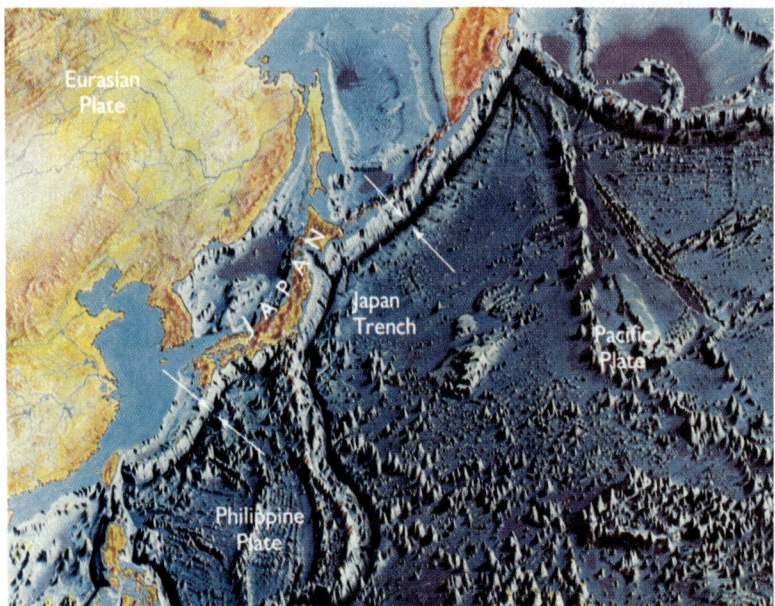

Seafloor topography of the Western Pacific shows the Japan Trench, part of the system of subduction zones that bound the Pacific, Philippine, and Eurasian plates. (World Ocean Floor, based on bathymetric studies by Bruce C. Heezen and Marie Tharp. Painting by Heinrich C. Berann. Copyright © Marie Tharp, 1977.)

SUMMARY

How are igneous rocks classified? All igneous rocks can be divided into two broad textural classes: (1) the coarsely crystalline rocks, which are intrusive and therefore cooled slowly, and (2) the finely crystalline ones, which are extrusive and cooled rapidly. Within each of these broad categories, the rocks are classified on the chemical basis of their silica content or by the mineralogical equivalent, the proportions of lighter, felsic minerals and darker, mafic minerals. Felsic rocks, such as granite and its corresponding extrusive, rhyolite, are rich in silica and dominated by quartz, potassium feldspar, and sodium-rich plagioclase feldspar. Mafic rocks, such as gabbro and its corresponding extrusive, basalt, are poor in silica and consist primarily of pyroxene, olivine, and calcium-rich feldspar. Intermediate rocks are granodiorite and diorite and their corresponding extrusives, dacite and andesite.

How and where do magmas form? Magmas form at places in the lower crust and mantle where temperatures and pressures are high enough for at least partial melting of water-containing rock. Basalt can partially melt in the upper mantle where convection currents bring hot rock upward at mid-ocean ridges. Mixtures of basalt and other igneous rocks with sedimentary rocks, which contain significant quantities of water, have lower melting points than dry igneous rocks. These mixtures therefore melt when they are heated during subduction into the mantle.

How does magmatic differentiation account for the variety of igneous rocks? Minerals crystallize from magmas along two paths: (1) a continuous reaction series of the plagioclase feldspars and (2) a discontinuous reaction series of the mafic minerals. In these series, crystals continuously react with the melt through successive stages of crystallization and magma composition until they solidify completely, at which point the final product has the same composition as the original magma. If there is fractional crystallization, so that the crystals do not react with the melt, either because they grow very rapidly or because they are separated from the liquid, the final product may be more silicic than the earlier, more mafic crystals.

Bowen's continuous and discontinuous reaction series explain how fractional crystallization can produce mafic igneous rocks from earlier stages of crystallization and differentiation and felsic rocks from later stages, but Bowen's theory does not adequately explain the abundance of granite. Magmatic differentiation of basalt does not explain the composition and abundance of igneous rocks. Different kinds of igneous rocks may be produced by variations in the compositions of magmas caused by the melting of different mixtures of sedimentary and other rocks and by mixing of magmas.

What are the forms of intrusive and extrusive igneous rocks? Igneous bodies of large size are plutons. The largest plutons are batholiths, which are thick tabular masses with a central funnel. Stocks are smaller plutons. Less massive than plutons are sills, which are concordant, with the intruded rock, following its layering, and dikes, which are discordant with the layering, cutting across it. Hydrothermal veins are formed where water is abundant, either in the magma or in surrounding country rock.

How are igneous rocks related to plate tectonics? The two major sites of magmatic activity are mid-ocean ridges, where basalt wells up from the upper mantle, and subduction zones, where a series of differentiated magmas produces both extrusives and intrusives in island or continental volcanic arcs as the subducting oceanic lithosphere moves down into the deep crust and upper mantle. Large volumes of basalt are produced at oceanic islands and on landmasses that overlie mantle plumes.

KEY TERMS AND CONCEPTS

igneous rocks (p. 75)
intrusive igneous rock (p. 78)
country rock (p. 78)
extrusive igneous rock (p. 78)
pyroclast (p. 78)
volcanic ash (p. 78)
tuff (p. 78)
pumice (p. 78)
obsidian (p. 78)
porphyry (p. 79)
felsic rocks (p. 81)
granite (p. 81)
rhyolite (p. 81)
intermediate igneous rocks (p. 81)
granodiorite (p. 81)

diorite (p. 81)
dacite (p. 81)
andesite (p. 81)
mafic rocks (p. 81)
gabbro (p. 82)
basalt (p. 82)
ultramafic rocks (p. 82)
peridotite (p. 82)
viscosity (p. 82)
partial melting (p. 83)
partial melt (p. 83)
magma chamber (p. 84)
magmatic differentiation (p. 87)
continuous reaction series (p. 88)
discontinuous reaction series (p. 88)

zoned crystal (p. 89)
fractional crystallization (p. 90)
Bowen reaction series (p. 92)
pluton (p. 95)
granitization (p. 96)
batholith (p. 96)
stock (p. 96)
discordant intrusion (p. 96)
sill (p. 96)
concordant intrusion (p. 96)
dike (p. 96)
vein (p. 98)
pegmatite (p. 98)
hydrothermal vein (p. 98)

EXERCISES

1. Why are intrusive igneous rocks coarsely crystalline and extrusive rocks finely crystalline?

2. What kinds of minerals would you find in a mafic igneous rock?

3. What kinds of igneous rock contain quartz?

4. Name two intrusive igneous rocks with a higher silica content than that of gabbro.

5. What is the difference between the continuous and discontinuous reaction series?

6. How does fractional crystallization lead to magmatic differentiation?

7. Where in the crust, mantle, or core would you find a partial melt of basaltic composition?

8. How do you distinguish a sill from a dike?

9. What field geological evidence could you use to tell whether a basalt formation was a dike or a lava flow?

10. In which plate-tectonic settings would you expect magmas to form?

11. Why do melts migrate upward?

12. Where on the ocean floor would you find basaltic magmas being extruded?

THOUGHT QUESTIONS

1. How would you classify a coarse-grained igneous rock that contained about 12 percent quartz, 10 percent potassium feldspar, 35 percent plagioclase feldspar, and small amounts of biotite and amphibole?

2. What kind of rock would contain some plagioclase feldspar crystals about 5 mm long "floating" in a dark-gray matrix of crystals of less than 1 mm?

3. What differences in crystal size might you expect to find between two sills, one intruded at a depth of about 12 km, where the country rock was very hot, and the other at a depth of 0.5 km, where the country rock was moderately warm?

4. If you were to drill a hole through the crust of a volcanic island arc, what intrusive or extrusive igneous rocks might you expect to encounter at or near the surface? What intrusive or extrusive igneous rocks might you expect at the base of the crust?

5. Assume that a magma with a certain ratio of calcium to sodium starts to crystallize. If fractional crystallization occurs during the solidification process, will the plagioclase feldspars formed after complete crystallization have the same ratio of calcium to sodium that characterized the magma?

6. Would you expect to find a magma crystallizing olivine at the same stage of crystallization as a sodium-rich plagioclase feldspar? Why, or why not?

7. In what way is a zoned crystal evidence of fractional crystallization?

8. Why are plutons more likely than dikes to show the effects of magmatic crystallization?

9. What might be the origin of a rock composed almost entirely of olivine?

10. Are porphyries more likely to occur in rapidly cooled extrusive rocks or in very slowly cooled intrusive rocks?

11. What kinds of rock would you expect to find if you drilled a hole in a mid-ocean ridge?

12. Water is abundant in the sedimentary rocks and oceanic crust of subduction zones. How would the water affect melting in these zones?

Suggested Readings

Barker, D. S. 1983. *Igneous Rocks.* Englewood Cliffs, N.J.: Prentice Hall.

Best, M. G. 1982. *Igneous and Metamorphic Petrology.* San Francisco: W. H. Freeman.

Blatt, Harvey, and Robert J. Tracy. 1996. *Petrology: Igneous, Sedimentary, and Metamorphic,* 2nd ed. New York: W. H. Freeman.

Coffin, Millard F., and Olav Eldholm. 1993. Large igneous provinces. *Scientific American* (June), pp. 42–49.

Philpotts, Anthony R. 1990. *Igneous and Metamorphic Petrology.* Englewood Cliffs, N.J.: Prentice Hall.

Raymond, L. A. 1995. *Petrology.* Dubuque, Iowa: Wm. C. Brown.

Internet Sources

Igneous Rocks
🛈 http://www.science.ubc.ca/~geol202/igneous/igneous.html
This URL for the University of British Columbia's Introduction to Petrology course (Geology 202) goes directly to the section on igneous rocks.

Granite
🛈 http://uts.cc.utexas.edu/~rmr/general.html#links
Established by a graduate student at the University of Texas, this site includes abundant information about granite (especially in Texas) and links to a variety of other sites.

5

Lava stream flowing from a vent near the summit of Mount Etna, Sicily, in the eruption of January 1992. *(Roger Ressmeyer/Starlight.)*

Volcanism

Imagine a volcanic eruption in which an area of ground the size of New York City collapses, a region larger than Vermont is buried under hot ash that snuffs out all life, and fields 1000 or 2000 km away are blanketed by 20 cm of ash and rendered infertile. Imagine that the volcanic dust thrown high into the stratosphere dims the Sun for a year or two, so that there are no summers. Unbelievable? Yet it has happened at least twice in what is now the United States: at Yellowstone in Wyoming 600,000 years ago and at Long Valley, California, 700,000 years ago. This was long before humans first reached North America, only 30,000 years ago, but not very long ago on the 4.5-billion-year geologic time scale. We know that these events took place because volcanic rocks formed by the eruptions have been identified.

About 80 percent of Earth's crust, oceanic as well as continental, is made up of volcanic rock.

Beginning as magma deep inside the Earth, volcanic rock serves, in a sense, as a window through which we can dimly perceive Earth's interior. This chapter examines **volcanism,** the process by which magma from the interior of the Earth rises through the crust, emerges onto the surface as lava, and cools into hard, volcanic rock. We will look at the major types of lava, eruptive styles, the landforms they create, and the kinds of environmental disruption that volcanoes can cause. We will see how plate tectonics can explain why the majority of volcanoes occur at plate boundaries but a few occur at "hot spots" within plates. Finally, we will discuss ways to control the destructive potential of volcanoes and benefit from their chemical riches and heat energy.

Ancient philosophers were awed by volcanoes and their fearsome eruptions of molten rock. In their efforts to explain volcanoes, they spun myths about a hellish, hot underworld below Earth's surface. The ancients had the right idea. Modern scientists, also seeking an explanation, see in volcanoes evidence of Earth's internal heat.

Temperature readings of rocks as far down as humans have drilled (about 10 km) show that the Earth does get hotter with depth. Inferring trends to even greater depths, geologists now believe that temperatures at the depths of the asthenosphere, which extends from about 75 to 250 km, reach 1100 to 1200°C—high enough for the rocks there to begin to melt. This is why geologists identify the asthenosphere as a main source of magma, the molten rock below Earth's surface that we call **lava** after it erupts. Melting of sections of the solid lithosphere that rides above the asthenosphere may be another source. The magma rises buoyantly, that is, it floats up because the fraction that melts at these temperatures is less dense than the residual surrounding rocks. In effect, the denser surrounding rock exerts pressure on the melt squeezing it upward. In some places the melt may find a path to the surface through fractures in the lithosphere. In other places geologists believe the rising magma melts a path to the surface. Some of the magma eventually reaches the surface and erupts as lava. A **volcano** is a hill or mountain that forms from the accumulation of matter that erupts at the surface. Figure 5.1 is a simplified, generic diagram of the plumbing system of a volcano, which taps a pool of molten rock at depth and gives vent to it at the surface. Note the pipelike conduit through which the magma rises from the magma chamber. This shallow reservoir in the crust below the summit periodically fills with magma rising from below and empties to the surface in cycles of eruptions. Lava can also erupt from cracks on the flanks of the volcano.

Because lava is a sample of the Earth's interior, it is interesting to geologists. Unfortunately, it is not a perfect sample. Lava differs from magma at depth. It will have lost some gaseous constituents to the atmosphere or ocean as it erupted and may have gained or lost other chemical components on its way to the surface. Despite these differences, lava and other eruptive materials still provide us with important information clues to the chemical com-

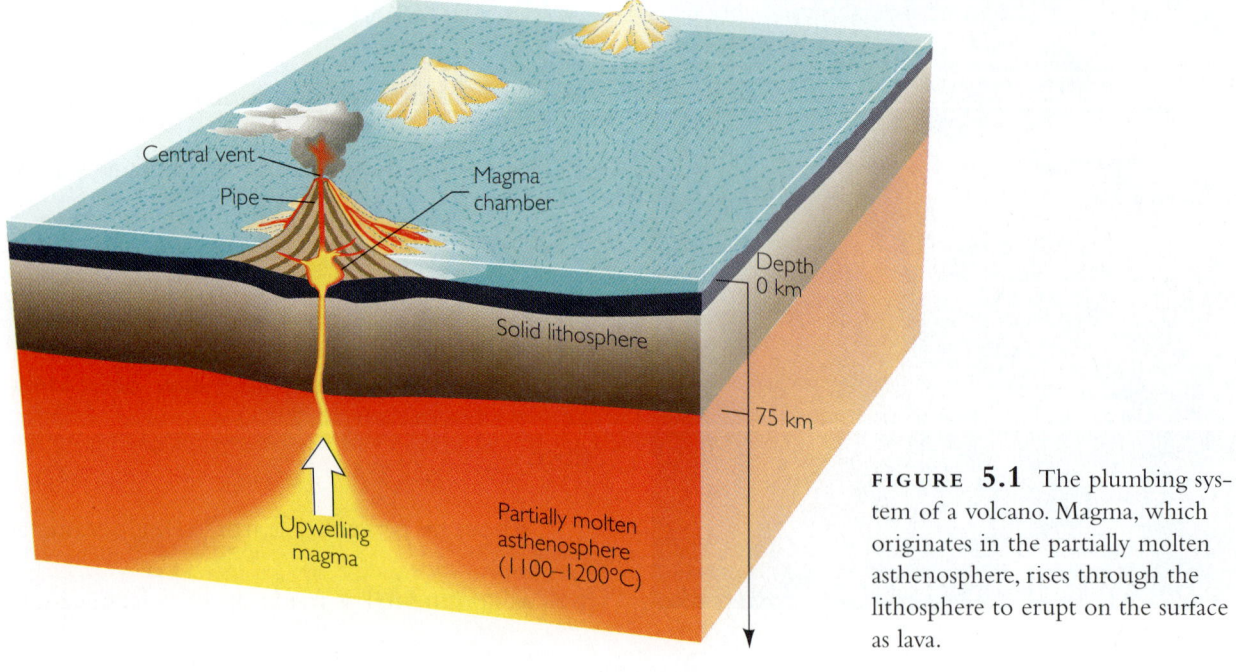

FIGURE **5.1** The plumbing system of a volcano. Magma, which originates in the partially molten asthenosphere, rises through the lithosphere to erupt on the surface as lava.

position and physical state of the upper mantle. Hardened into volcanic rock, these materials can also tell us something about eruptions that created them many thousands or millions of years ago. The chemical and mineralogical compositions of lava have much to do with the way it erupts and the kind of landform it will create when it solidifies.

VOLCANIC DEPOSITS

The major types of lavas and the rocks they form differ according to the magmas from which they were derived. As we saw in Chapter 4, igneous rocks and their precursor magmas are divided into three major groups—felsic, intermediate, and mafic—based on their chemical composition (see Table 4.2). The rocks are further classified as intrusive (they cooled below the surface and are coarse-grained as a result) or extrusive (they cooled on the surface and are finer-grained). The major intrusive rocks are granite (felsic), diorite (intermediate), and gabbro (mafic). The major extrusive counterparts are rhyolite (felsic) and the more common andesite (intermediate) and basalt (mafic). These classifications are summarized in Figure 4.6. With this framework in mind, let us examine the types of lava and how they flow and solidify.

Types of Lava

The several types of lavas leave behind different landforms—volcanic mountains that vary in shape and solidified lava flows that vary in character. These variations reflect differences in the chemical composition, gas content, and temperature of lavas. The higher the silica content and the lower the temperature, for example, the more viscous the lava is and the more slowly it flows. The more gas a lava contains, the more violent its eruption is likely to be.

BASALTIC LAVAS Basaltic lava, dark in color, erupts at temperatures of 1000 to 1200°C—close to the temperature of the upper mantle. Because of its high temperature and low silica content, basaltic lava is extremely fluid and can flow downhill fast and far. Streams flowing as fast as 100 km per hour have been observed, although velocities of a few kilometers per hour are more common. In 1938 two daring Russian volcanologists measured temperatures and collected gas samples while they floated down a river of molten basalt on a raft of colder solidified lava. The surface temperature of their raft was 300°C, and the river had a temperature of 870°C. Lava streams that traveled more than 50 km from their source have been witnessed in historical times.

Basaltic lava flows vary according to the conditions under which they erupt. Important examples include:

- *Flood Basalts* Highly fluid basaltic lava that erupts on flat terrain, however, can spread out in thin sheets as a flood of lava. Successive flows often pile up into immense basaltic lava plateaus, called flood basalts, as at the great Columbia Plateau of Oregon and Washington (Figure 5.2).

- *Pahoehoe and Aa* Cooling basaltic lava flowing downhill falls into one of two categories,

FIGURE 5.2 View of Columbia Plateau, Washington. Successive flows of flood basalts piled up to build this immense plateau, which covers a large area of Washington and Oregon. *(Martin G. Miller.)*

FIGURE 5.3 Two types of lava, ropy pahoehoe (*bottom*) and jagged blocks of aa (*top*). Mauna Loa volcano, Hawaii. *(Kim Heacox/DRK.)*

according to its surface form: **pahoehoe** (pronounced pa-ho-ee-ho-ee) or **aa** (ah-ah). Figure 5.3 shows examples of both.

Pahoehoe (the word is Hawaiian for "ropy") forms when a highly fluid lava spreads in sheets and a thin, glassy elastic skin congeals on its surface as it cools. The skin is dragged and twisted into coiled folds that resemble rope as the molten liquid continues to flow below the surface (see Figure 5.3).

"Aa" is what the unwary exclaim after venturing barefoot onto lava that looks like clumps of moist, freshly plowed earth. Aa is lava that has lost its gases and consequently become more viscous than pahoehoe. It moves more slowly, allowing a thick skin to form. As the flow continues to move, the thick skin breaks into rough, jagged blocks (see Figure 5.3). The blocks ride on the viscous, massive interior, piling up a steep front of angular boulders that advances like a tractor tread. Aa is truly treacherous to cross. A good pair of boots may last about a week on it, and the traveler or geologist can count on cut knees and elbows.

A single downhill basaltic flow will commonly have the features of pahoehoe near its source, where the lava is still fluid and hot, and of aa farther downstream, where the flow's surface, having been exposed to cool air for more time, has developed a thicker outer layer.

- *Pillow Lavas* A geologist who comes across **pillow lavas**—piles of ellipsoidal, sacklike blocks of basalt about a meter wide—knows they formed in an underwater eruption (Figure 5.4) even if they are now on dry land. In fact, pillow lavas are an important indicator that a region was once under water. Geologist-divers have actually observed pillow lavas forming on the ocean floor off Hawaii. Tongues of molten basaltic lava develop a tough, plastic skin on contact with the cold ocean water. Because lava inside the skin cools more slowly, the pillow's interior develops a crystalline texture, while the quickly chilled skin solidifies to a crystalless glass.

RHYOLITIC LAVAS Rhyolite, light in color and the most felsic lava, has a lower melting point than basalt and erupts at temperatures of 800 to 1000°C. It is much more viscous because of its lower temperature and higher silica content. Rhyolite moves 10 or more times more slowly than basalt, and because it resists flow, it tends to pile up in thick, bulbous deposits.

ANDESITE LAVAS Andesite, with an intermediate silica content, has properties that fall between those of basalts and rhyolites.

FIGURE 5.4 Pillow lava, characteristic of underwater volcanic eruptions, on the seafloor near the Galápagos Islands. Pillow lava found on land indicates earlier period of submergence. *(Woods Hole Oceanographic Institute.)*

Textures of Lavas

Lavas have other features that reflect the temperature and pressure conditions under which they formed. They can have a glassy or fine-grained texture if they cool quickly or a coarsely crystalline texture if they cool slowly beneath the surface. They can also have little bubbles, created when pressure falls suddenly as the lava rises and cools. Lava is typically charged with gas, like soda in an unopened bottle. When lava rises, the pressure on it decreases, just as the pressure on the soda drops when the bottle cap is removed. And just as the soda's carbon dioxide creates bubbles as it is released, water vapor and other dissolved gases escaping from lava create gas cavities, or *vesicles* (Figure 5.5). A frothy texture in solidified lava provides geologists with details of the rock's volcanic origins. One extremely vesicular, generally rhyolitic volcanic rock is pumice. Some pumice has so much void space that it is light enough to float.

Pyroclastic Deposits

Water and gases in magmas can have even more dramatic effects on eruptive styles. Before eruption, the confining pressure of the overlying rock keeps these volatiles from escaping. When the magma rises close to the surface and the pressure drops, the volatiles may be released with explosive force, shattering the lava and any overlying solidified rock into fragments of various sizes, shapes, and textures (Figure 5.6). Explosive eruptions are particularly likely with gas-rich, viscous rhyolitic and andesitic lavas.

VOLCANIC EJECTA Pyroclasts, as described in Chapter 4, are any fragmentary volcanic rock materials that are ejected into the air. These rocks, miner-

FIGURE 5.6 Pyroclastic eruption at Arenal volcano, Costa Rica. *(Gregory G. Dimijian/Photo Researchers.)*

als, and glasses are classified according to size. The finest fragments, less than 2 mm in diameter, are called **ash.** Fragments ejected as blobs of lava that become rounded and cooled in flight or chunks torn loose from previously solidified volcanic rock can be much larger (Figure 5.7). Volcanic ejecta as big as a house have been thrown more than 10 km in violent eruptions. Volcanic ash fine enough to stay

FIGURE 5.5 Vesicular basalt sample, approximately 1 ft across. *(Glenn Oliver/Visuals Unlimited.)*

FIGURE 5.7 Volcanologist Katia Krafft examines a volcanic bomb ejected from Asama volcano, Japan. *(Science Source/Photo Researchers.)*

FIGURE 5.8 Volcanic breccia. The view covers about 1 ft across. *(Doug Sokell/Visuals Unlimited.)*

aloft can be carried great distances. Within two weeks of the 1991 eruption of Mount Pinatubo in the Philippines, its volcanic dust was traced all the way around the world by orbiting Earth satellites.

Sooner or later pyroclasts fall, usually building deposits near their source. As they cool, hot sticky fragments become welded together (or lithified). The rocks created from smaller fragments are called **volcanic tuffs;** those formed from larger fragments are called **volcanic breccias** (Figure 5.8).

PYROCLASTIC FLOWS One particularly spectacular and often devastating form of eruption occurs when hot ash, dust, and gases are ejected in a glowing cloud that rolls downhill at speeds of up to 200 km per hour. The solid particles are actually buoyed up by hot gases, so there is little frictional resistance to this incandescent **pyroclastic flow** (Figure 5.9).

In 1902 a pyroclastic flow with an internal temperature of 800°C exploded from the side of Mont Pelée, on the Caribbean island of Martinique, with very little warning. The avalanche of choking hot gas and glowing volcanic ash plunged down the slopes at a hurricane speed of 160 km per hour. In one minute and with hardly a sound, the searing emulsion of gas, ash, and dust enveloped the town of St. Pierre and killed 29,000 people. It is sobering to scientists who render advice to others to recall the statement of one Professor Landes, issued the day before the cataclysm: "The Montagne Pelée presents no more danger to the inhabitants of St. Pierre than does Vesuvius to those of Naples." Professor Landes perished with the others. French volcanologists Maurice and Katia Krafft, whose photographs appear in this chapter, were killed by a pyroclastic flow at Mount Unzen, Japan, in 1991.

FIGURE 5.9 A pyroclastic flow plunges down the slopes of Mount Unzen, in Japan, in June 1991. Note the fireman and fire engine in the foreground, trying to outrun the hot ash cloud descending on them. Three scientists who were studying this volcano died when they were engulfed by a similar flow. *(AP/Wide World Photos.)*

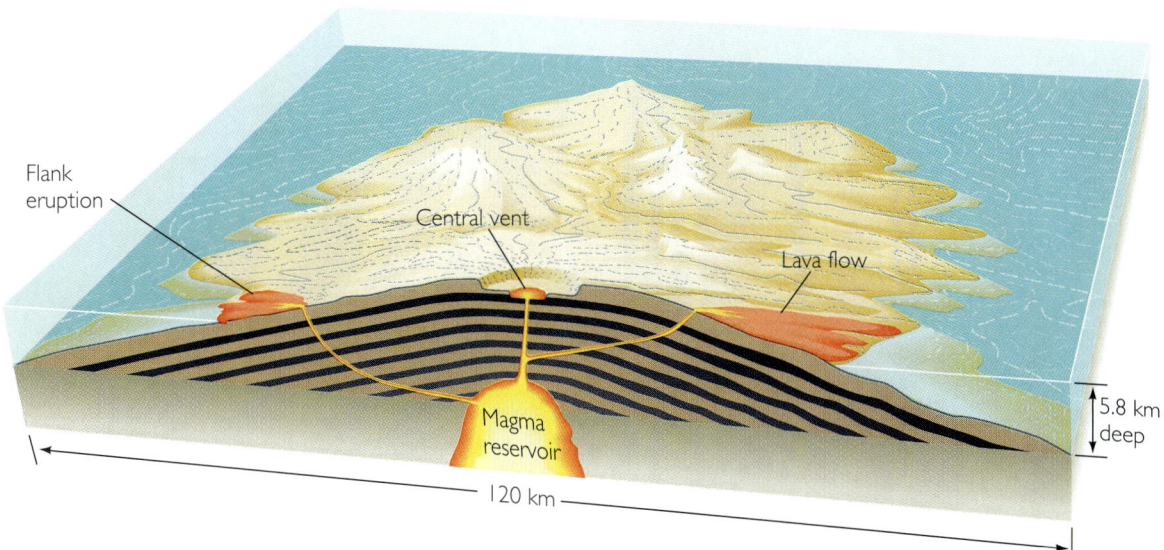

FIGURE 5.10 A shield volcano is built up by the accumulation of thousands of thin basaltic lava flows that spread widely and cool as gently sloping sheets. Each layer shown in this schematic diagram represents an accumulation of many hundreds of thin lava flows. Magma can find alternate routes to the surface and erupt on the flanks of a volcano. Modeled after Mauna Loa, Hawaii.

Eruptive Styles and Landforms

Now that we have examined various types of volcanic materials that pour or blow out of Earth's interior, we can look more closely at styles of eruptions and the characteristic formations they leave behind. Eruptions do not always create a majestically symmetrical cone. The hundreds of thousands of square kilometers of monotonous layers of basalt that make up the Columbia Plateau of Washington and Oregon present another variant. Volcanic landforms vary in shape with the properties of the lava and the conditions under which it erupts.

Central Eruptions

Central eruptions create the most familiar of all volcanic features—the volcanic mountain shaped like a cone. These eruptions discharge lava or pyroclastic materials from a **central vent,** an opening atop a pipelike feeder channel rising from the magma chamber through which the material rises to erupt at the Earth's surface.

SHIELD VOLCANOES   A lava cone is built by successive flows of lava from a central vent. If the lava is basaltic, it flows easily and spreads widely. If flows are copious and frequent, they create a broad, shield-shaped volcano many tens of kilometers in circumference and more than 2 km high. The slopes are relatively gentle. Mauna Loa, on Hawaii, is the classic example of a **shield volcano** (Figure 5.10). Although it rises only 4 km above sea level, it is actually the world's tallest structure: measured from its base on the seafloor, the volcano is 10 km high. It has a base diameter of 120 km—an area roughly three times that of Rhode Island. It grew to this enormous size by the accumulation of thousands of lava flows, each only a few meters thick, over a period of a few million years. In fact, the island of Hawaii actually consists of the tops of a series of overlapping active shield volcanoes emerging through the ocean surface.

VOLCANIC DOMES   In contrast to basaltic lavas, felsic lavas are so viscous that they can just barely flow. They usually produce a **volcanic dome,** a rounded, steep-sided mass of rock. Domes look as though lava had been squeezed out of a vent like toothpaste, with very little lateral spreading. Domes often plug vents, trapping gases. Pressures increase until an explosion occurs, blasting the dome into fragments. This occurred in the eruptions of Mount St. Helens in 1980 (Figure 5.11).

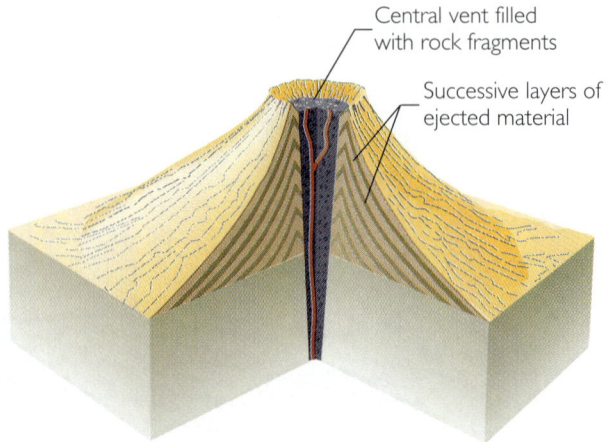

FIGURE 5.12 In a cinder cone, ejected material is deposited as layers that dip away from the crater at the summit. The vent beneath the crater is filled with fragmental debris.

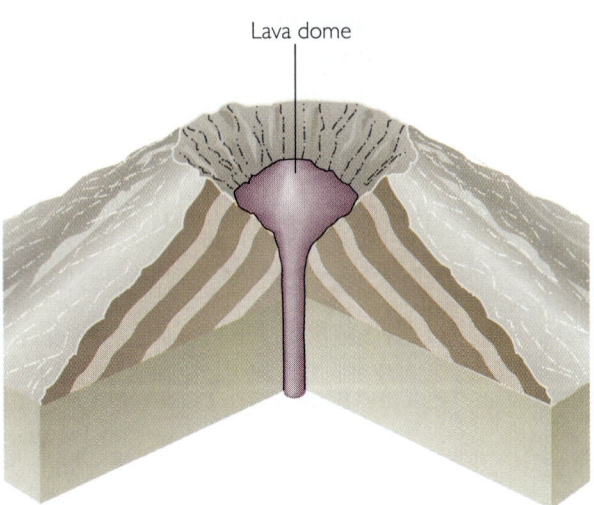

FIGURE 5.11 Volcanic domes are bulbous masses of felsic lava, which are so viscous that instead of flowing they pile up over the vent. Shown here is a growing dome within the crater of Mount St. Helens after the eruption. *(Lyn Topinka/USGS Cascades Volcano Observatory.)*

FIGURE 5.13 Cerro Negro in 1968. This volcano, near Managua, Nicaragua, is a cinder cone built on an older terrain of lava flows. *(Mark Hurd Aerial Surveys.)*

CINDER-CONE VOLCANOES When volcanic vents discharge pyroclasts, the solid fragments build up and form **cinder cones** (Figure 5.12). The profile of a cone is determined by the maximum angle at which the debris remains stable instead of sliding downhill (see Figure 11.1). The larger fragments, which fall near the summit, can form very steep but stable slopes. Finer particles are carried farther from the vent and form gentle slopes at the base of the cone. The classic concave-shaped volcanic cone with its summit vent reflects this variation in slope (Figure 5.13).

COMPOSITE VOLCANOES When a volcano emits lava as well as pyroclasts, alternating lava flows and beds of pyroclasts build a concave-shaped **composite volcano** or **stratovolcano** (Figure 5.14). This is the most common form of such large volcanoes as Fujiyama in Japan (Figure 5.15), Mounts Vesuvius and Etna in Italy, and Mount St. Helens in Washington.

CRATERS A bowl-shaped pit or **crater** is found at the summit of most volcanoes, centered on the vent. During the eruption of a lava volcano, the upwelling lava overflows the crater walls. When eruption ceases, the lava that remains in the crater often sinks back into the vent and solidifies. When the next eruption occurs, the material is literally blasted out of the crater in a pyroclastic explosion. The crater later becomes partially filled by the debris that falls back into it. Because a crater's walls are steep, they may cave in or become eroded in time. In this way the diameter of a crater can grow to several times that of the vent and hundreds of meters deep. The crater of Mount Etna in Sicily, for example, is presently 300 m (more than three football fields) in diameter and at least 850 m deep.

CALDERAS After a violent eruption in which large volumes of magma are discharged from a magma chamber a few kilometers below the vent, the empty chamber may no longer be able to support its roof. In such cases the overlying volcanic structure can

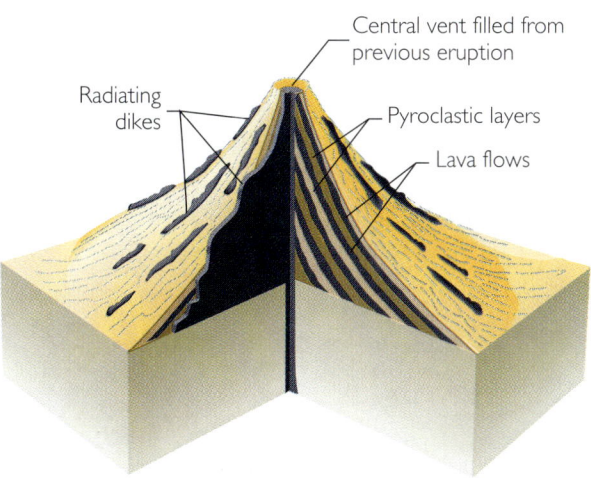

FIGURE 5.14 A composite volcano is built up of alternating layers of pyroclastic material and lava flows. Lava that has solidified in fissures forms riblike dikes that strengthen the cone. *(After R. G. Schmidt, USGS.)*

FIGURE 5.15 The composite volcano Fujiyama, Japan. *(Raga/The Stock Market.)*

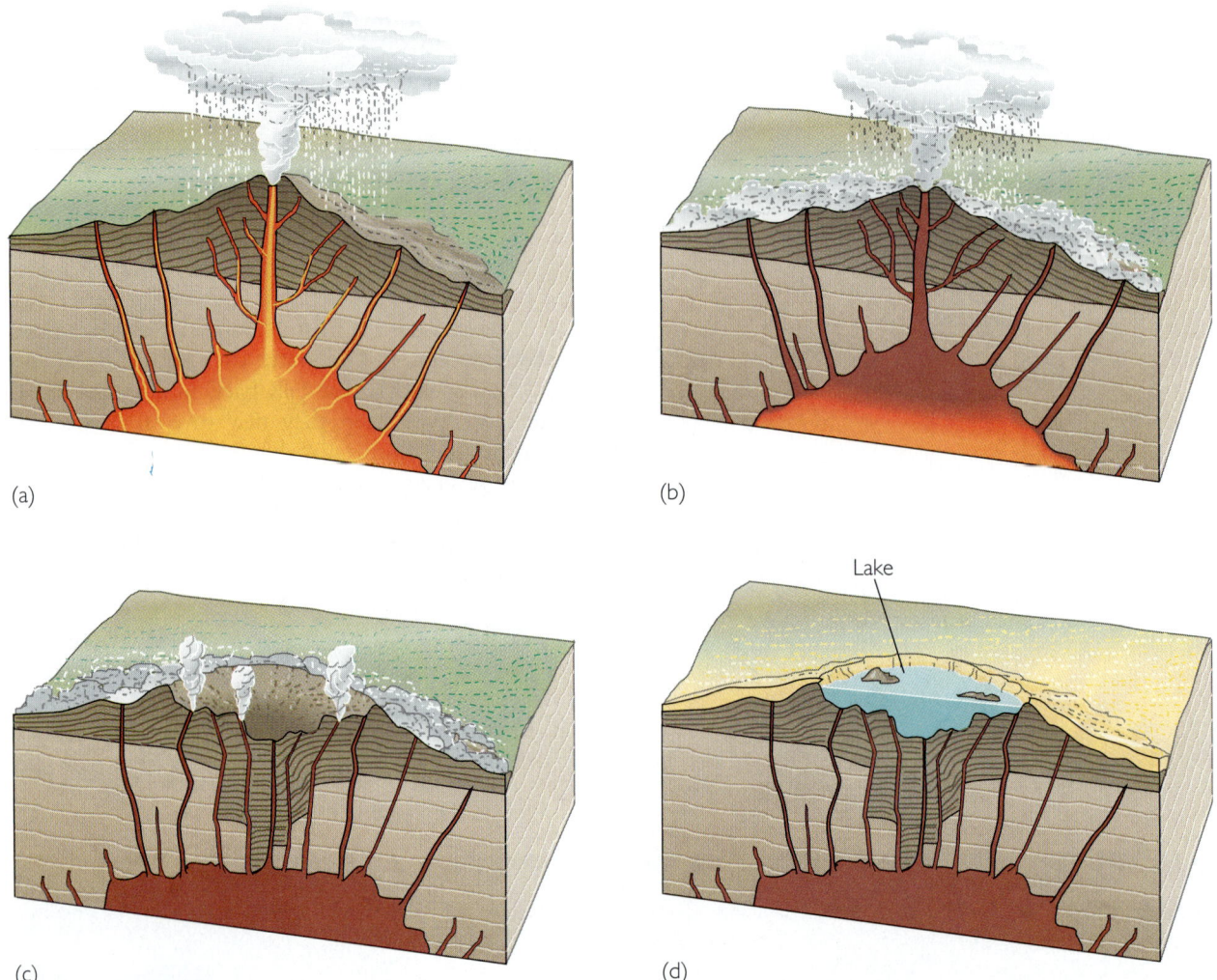

FIGURE 5.16 Stages in the evolution of a caldera. (a) Fresh magma fills a magma chamber and triggers a volcanic eruption of lava and columns of incandescent ash. (b) Eruption of lava and pyroclastic flows continue and the magma chamber becomes partially depleted. (c) Caldera results from the collapse of the mountain summit into empty chamber. Large pyroclastic flows ejected from fractures accompany the collapse, blanketing the caldera and the surrounding area. (d) A lake can form in the caldera. As the residual magma in the chamber cools, minor eruptive activity continues, with hot springs and gas emissions, and a small volcanic cone forms in the caldera.

collapse catastrophically, leaving a large, steep-walled, basin-shaped depression, much larger than the crater, called a **caldera** (Figure 5.16). Calderas are impressive features, ranging in size from a few kilometers to as much as 50 km or more in diameter, or about the area of greater New York City (Figure 5.17).

After some hundreds of thousands of years, fresh magma can reenter the collapsed magma chamber and reinflate it, forcing the caldera floor to dome upward again—perhaps to repeat the cycle of eruption, collapse, resurgence, eruption, and so on. This phenomenon is known as a *resurgent caldera*. The collapse of large resurgent calderas is one of the most destructive natural phenomena on Earth. Yellowstone Caldera in Wyoming, marked today by a leftover relic, Old Faithful geyser, ejected some 1000 km^3 of pyroclastic debris during its eruptive stages about 600,000 years ago, more than a thousand times the amount of material ejected by Mount St. Helens in 1980. Ash deposits fell over much of what is now the United States. Other resurgent calderas are Valles Caldera in New Mexico, Long Valley Caldera in California, the still-active Kilauea and Rabaul (New Guinea) calderas, and the dormant Crater Lake in Oregon.

FIGURE 5.17 Crater Lake, Oregon, fills a caldera 8 km in diameter. The caldera is all that remains of an earlier composite volcano that was destroyed in the collapse that formed the caldera. *(Greg Vaughn/ Tom Stack.)*

Monitoring caldera unrest is very important to geologists today because of the potential for large destruction. Fortunately, no catastrophic collapses with global consequences have occurred during human history, but geologists are wary of an increasing occurrence of small earthquakes in Yellowstone and Long Valley calderas and other indications of activity in the magma chambers in the crust below. For example, carbon dioxide leaking into soil from magma deeper in the crust has been killing trees since 1992 on Mammoth Mountain, a volcano on the boundary of Long Valley Caldera. A forest service ranger was almost asphyxiated by volcanic carbon dioxide leaking into a cabin on the flanks of Mammoth Mountain. Other indications of resurgence of the Long Valley Caldera are the occurrence of nearly continuous swarms of small earthquakes and uplift of the center of the caldera by about half a meter over the past 15 years. Rabaul Caldera showed uplift of about 6 meters in the 20 years before violent eruptions occurred in 1994. Following the eruption the caldera subsided by about 2 meters.

At one time it was thought that a caldera was formed by a huge explosion in which a volcano literally blew its top. However, geologic mapping of the debris and the pattern of faulting around calderas produces a picture more consistent with the collapse of the roof than with its ejection upward.

PHREATIC EXPLOSIONS When hot, gas-charged magma encounters groundwater or seawater, the vast quantities of superheated steam generated cause **phreatic,** or steam, **explosions** (Figure 5.18). One of the most destructive volcanic eruptions in history, that of Krakatoa in Indonesia, was a phreatic explosion (see Feature 5.1).

FIGURE 5.18 Phreatic explosion on Nisino-sima, a new volcano that rose above the sea in 1973 following a submarine eruption in the Pacific Ocean about 900 km south of Tokyo. *(Maritime Safety Agency, Japan.)*

5.1 INTERPRETING THE EARTH

The Explosion of Krakatoa

The 1883 explosion of the volcano Krakatoa, in the strait between Java and Sumatra, was one of the greatest ever witnessed. Now almost completely submerged, Krakatoa was then a small island formed from a group of volcanic cones in an ancient caldera. The caldera, 6 km across, was a remnant of a collapsed prehistoric andesitic stratovolcano. On August 27, after many smaller explosions, Krakatoa blew its top in a phreatic explosion with the energy of 100 million tons of TNT (5000 times greater than the nuclear explosion that destroyed Hiroshima). It is believed that much of the energy was provided by the violent expansion of hot steam after the walls of the volcano first ruptured, letting seawater into the magma chamber. The result can be viewed as the biggest steam-boiler explosion and the loudest noise in recorded history.

The explosion was heard in Australia, nearly 2000 km away. Volcanic ash fell over an area of some 700,000 km^2. Almost total darkness settled on Jakarta, 150 km away, when the dust blotted out the Sun's rays. Fine dust rose to the stratosphere and drifted around the Earth, lowering Earth's mean annual temperature a few degrees for the next year or so by blocking 13 percent of the Sun's light from reaching Earth. The explosion also generated a *tsunami,* or giant sea wave, that reached a height of almost 40 m, destroying 295 coastal towns as far as 80 km away and drowning 36,000 people. The tsunami was recorded on tide gauges as far away as the English Channel. After the eruption, most of Krakatoa disappeared, leaving in its place the current 300-m-deep water-covered basin.

Anak Krakatoa ("Child of Krakatoa") is a new volcano that is building up in the caldera left by the 1883 eruption. The new cone rose above sea level in 1928. Since then, successive eruptions have largely filled in the 1883 caldera and may eventually build a new island in roughly the same position as the original Krakatoa. (*Katia Krafft/Explorer.*)

DIATREMES Sometimes when hot matter from the deep interior escapes explosively, the vent and feeder channel below are left filled with breccia as the eruption wanes. The resulting structure is called a **diatreme.** Shiprock, which towers over the surrounding plain in New Mexico, is a diatreme exposed by the erosion of the sedimentary rocks through which it originally burst. To transcontinental air travelers, Shiprock looks like a gigantic black skyscraper in the red desert (Figure 5.19).

The eruptive mechanism that produces diatremes has been pieced together in great detail from the geologic record. The kinds of minerals and rocks found in some diatremes could have been formed only at great depths—100 km or so, well within the upper mantle. This observation indicates that diatremes are formed when gas-charged magmas melt their way upward, finally ejecting gases, lava fragments from the vent walls, and fragments from the deep crust and mantle, all with explosive energy and sometimes at supersonic speed. Such an eruption would probably look like the exhaust jet of a giant rocket upside down in the ground blowing rocks and gases into the air.

Another diatreme is encountered in the underground workings of the fabled Kimberly mines of South Africa, one of the world's richest sources of diamonds. This diatreme is a peridotite, an ultramafic rock composed mostly of the mineral olivine. It also contains diamonds, which form from carbon under the great pressures found in the mantle, and other scrambled fragments of mantle rock picked up by the magma en route to the surface. Geologists view this diatreme much as they would a 300-km drill core into the mantle. Its fragments provide our only direct evidence of mantle materials and so have been studied extensively. This is an example of the evidence that enables geologists to conclude that peridotite is a major constituent of the upper mantle.

Fissure Eruptions

Imagine basaltic lava flowing out of a crack in the Earth's surface tens of kilometers long, flooding vast areas. Such **fissure eruptions** have occurred on Earth countless times in the last four billion years (Figure 5.20). Among the geologically important fissure eruptions are those that occur along mid-ocean ridges. In recorded history humans have witnessed such an eruption only once, in 1753 on Iceland, which is an exposed segment of the Mid-Atlantic Ridge. One-fifth of the Icelandic population perished as a result. A fissure 32 km long opened and

FIGURE 5.19 Shiprock, towering 515 m above the surrounding flat-lying sediments of New Mexico, is a diatreme, or volcanic pipe, exposed by erosion of its enclosing sedimentary rock. *(Fred Padula.)*

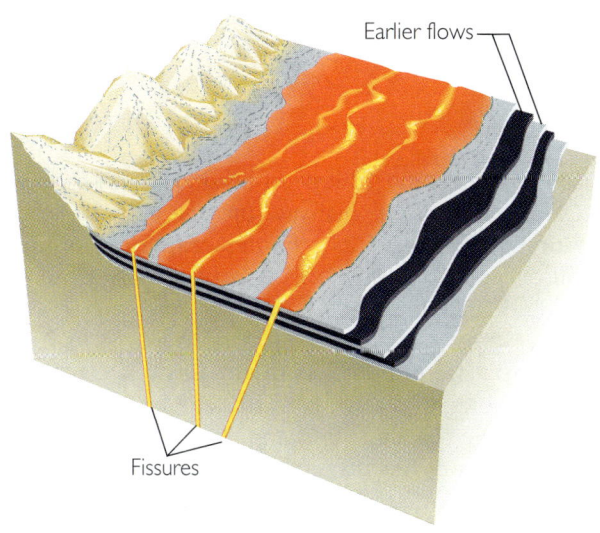

FIGURE 5.20 In a fissure eruption of highly fluid basalt, lava rapidly flows away from fissures and forms widespread layers, rather than building up into a volcanic mountain. *(After R. S. Fiske, USGS.)*

FIGURE 5.21 Volcanic cones along the Laki fissure (Iceland) that opened in 1783 and erupted the largest flow of lava on land in the course of human history. (*Tony Waltham.*)

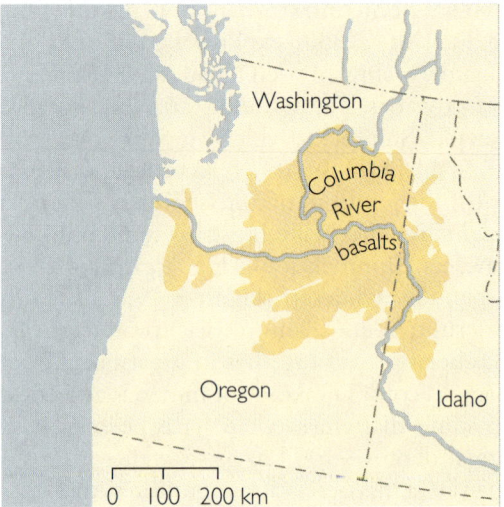

FIGURE 5.22 The area covered by the Columbia River flood basalts. (*After R. S. Fiske, USGS.*)

spewed out some 12 km^3 of basalt, enough to cover Manhattan about halfway up the Empire State Building (Figure 5.21). Fissure eruptions continue on Iceland, although on a smaller scale than the 1783 catastrophe.

Flood Basalts (Basaltic Lava Plateaus)

The geologic record contains ample evidence of prehistoric basaltic flooding from great fissures. When **flood basalts** erupt from fissures, the lavas build a plain or accumulate as a plateau, rather than piling up as a volcanic mountain as they do when they erupt from a central vent. The flood basalts that made the Columbia Plateau (see Figure 5.2) buried 200,000 km^2 of preexisting topography (Figure 5.22). Some individual flows were more than 100 m thick, and some were so fluid that they spread more than 60 km from their source. An entirely new landscape with new river valleys has since evolved atop the lava that buried the old surface. Plateaus made by flood basalts are found on every continent.

Ash-Flow Deposits Fissure eruptions of pyroclastic materials have produced extensive sheets of hard volcanic tuffs called **ash-flow deposits** (Figures 5.23 and 5.24). As far as is known, humans have never witnessed one of these spectacular events. The early Tertiary ash-flow deposits of the Great Basin in Nevada and adjacent states, formed in this way, cover an area of about 200,000 km^2 and are as much as 2500 m thick in some places. Yellowstone National

FIGURE 5.23 Welded tuff from an ash-flow deposit in the Sierra Nevada of California. Image covers an area of 2 ft × 3 ft. (*Gerald and Buff Corsi/Visuals Unlimited.*)

FIGURE 5.24 These ash-flow sheets on O-shima Island, Japan, were formed when intensely hot gas-charged volcanic dust, ash, and pumice spread swiftly over the surface to settle, cool, and harden. (*S. Aramaki.*)

Park, in Wyoming, has been covered by a number of ash-flow sheets (see Figure 5.24). A succession of forests there were buried by these deposits.

Other Volcanic Phenomena

LAHARS Among the most dangerous volcanic events are the torrential mudflows of wet volcanic debris that are called **lahars.** They can form when a pyroclastic flow meets a river or a snowbank; when the wall of a crater lake breaks, suddenly releasing water; when glacial ice is melted by a lava flow; or when heavy rainfall transforms new ash deposits into mudflows. One extensive layer of volcanic debris in the Sierra Nevada of California contains 8000 km^3 of material of lahar origin, enough to cover all of Delaware with a deposit more than a kilometer thick. Lahars have been known to carry huge boulders for tens of kilometers. When Nevado del Ruiz in the Colombian Andes erupted in 1985, lahars triggered by the melting of glacial ice near the summit plunged down the slopes and buried the town of Armero 50 km away, killing more than 25,000 people.

EDIFICE COLLAPSE The cone or edifice of a volcano is constructed from thousands of deposits of lava or ash or both. This is not the way to build a solid structure. In recent years volcanologists have discovered many examples of catastrophic structural failures in which a big piece of the summit breaks off, perhaps precipitated by an earthquake, and slides downhill in a massive, destructive landslide. Edifice failure occurs about four times a century. The collapse of one side of Mount St. Helens was the most damaging part of its 1980 eruption (look ahead to the photos in Feature 5.3, page 128). Surveys of the seafloor of Hawaii have discovered many giant landslides on the underwater flanks of the Hawaiian ridge. When they occurred, these massive movements would have triggered huge tsunami. Actually coral-bearing marine sediments have been found on one of the Hawaiian Islands some 300 m above sea level. These sediments were probably deposited by giant sea waves that were excited by a prehistoric volcanic collapse. It is worrisome that the southern flank of Kilauea volcano is advancing toward the sea at a rate of 25 cm per year. Should it break off and slide catastrophically into the ocean, it would prove disastrous for Hawaii, California, and all Pacific coastal areas.

VOLCANIC GASES The nature and origin of volcanic gases are of considerable interest and importance. It is thought that over geologic time these gases have created the oceans and the atmosphere and may even affect today's climate. Volcanic gases have been collected by courageous volcanologists and analyzed to determine their composition. Water vapor is the main constituent of volcanic gas (70 to 95 percent), followed by carbon dioxide, sulfur dioxide, and traces of nitrogen, hydrogen, carbon

FIGURE 5.25
Volcanologist Katia Krafft in a heatproof suit examining a lava flow on Kilauea volcano, Hawaii. *(Maurice Krafft/Photo Researchers.)*

monoxide, sulfur, and chlorine. Every eruption releases enormous amounts of these gases (Figure 5.25). Some volcanic gas may come from deep within the Earth, making its way to the surface for the first time. Some may be recycled groundwater and ocean water, recycled atmospheric gas, or gas that has been trapped in earlier generations of rocks.

The relation between volcanic eruptions and changes in weather and climate is receiving increasing attention. For instance, the 1982 eruption of El Chichón in southern Mexico and the 1991 eruption of Mount Pinatubo injected sulfurous gases into the atmosphere, 10 km above the Earth. Through various chemical reactions the gases formed an aerosol (a fine airborne mist) representing tens of millions of metric tons of sulfuric acid. This aerosol partially blocked enough of the Sun's radiation from reaching Earth's surface to lower global temperatures for a year or two. The eruption of Mount Pinatubo, one of the largest explosive eruptions of the century, led to a global cooling of at least 0.5°C in 1992. For similar reasons, the debris lofted into the atmosphere during the 1815 eruption of Mount Tambora in Indonesia caused even greater cooling. The Northern Hemisphere suffered a very cold summer with snow storms in 1816. The drop in temperature and ash fallout caused widespread crop failures. More than 90,000 people perished as a result in that "year without a summer." This terrible year inspired Byron's gloomy poem "Darkness":

I had a dream, which was not all a dream.
The bright sun was extinguish'd, and the stars
Did wander darkling in the eternal space,
Rayless, and pathless, and the icy earth
Swung blind and blackening in the moonless air;
Morn came and went—and came, and brought
  no day.
And men forgot their passions in the dread
Of this their desolation; and all hearts
Were chill'd into a selfish prayer for light . . .

Chlorine emissions from Pinatubo have also hastened the loss of ozone in the atmosphere, nature's shield that protects humankind from the Sun's ultraviolet radiation.

FUMAROLES, HOT SPRINGS, AND GEYSERS
Volcanic activity does not cease when lava or pyroclastic materials cease to flow. For decades and in some cases centuries after a major eruption, volcanoes continue to emit gas fumes and steam through small vents called **fumaroles.** All of these emanations contain dissolved materials that precipitate on surrounding surfaces as the water evaporates or cools. Various sorts of encrusting deposits (such as travertine) are formed, including some that contain valuable minerals (Figure 5.26).

Circulating groundwater that reaches buried magma (which retains heat for hundreds of thousands of years) is heated and returned to the surface

FIGURE 5.26 Fumarole becoming encrusted with sulfur deposits on Sierra Negra volcano, Galápagos Islands. *(Christian Grzimek/Photo Researchers.)*

FIGURE 5.27 Strokkur geyser, in Iceland, throws a column of steam and superheated water 20 to 30 m into the air every few minutes. *(Simon Fraser/Photo Researchers.)*

as hot springs and geysers (Figure 5.27). A geyser is a hot-water fountain that spouts intermittently with great force, frequently accompanied by a thunderous roar. The best known geyser in the United States is Old Faithful in Yellowstone Park, which erupts about every 65 minutes, sending a jet of hot water as high as 60 m.

THE GLOBAL PATTERN OF VOLCANISM

Before the advent of plate-tectonics theory, geologists noted a concentration of volcanoes around the rim of the Pacific Ocean and nicknamed it the "Ring of Fire." We now know that the Ring of Fire coincides with plate boundaries. This correlation has been an important clue to the origin of volcanoes.

The 500 to 600 active volcanoes of the world are not randomly distributed, but show a definite pattern. About 80 percent are found at boundaries where plates converge, 15 percent where plates separate, and the remaining few within plates (Figure 5.28). In addition, Chapter 4 shows that lava compositions vary with plate-tectonic setting (see Figure 4.8). Can we weave these observations into a hypothesis that not only describes events but explains them as well? Thanks to the theory of plate tectonics, we can.

Spreading-Zone Volcanism

As we saw in Chapters 1 and 4, the seafloor is broken up by a worldwide system of rifts, along which plates separate and basalt erupts. The fissure between the separating plates extends down to the asthenosphere. Basaltic magmas, partial melts of the hot, ultramafic rock in the asthenosphere (see pages 83–86 and Figure 4.8), rise buoyantly in the gap between the separating plates and overflow the fissure to form ocean ridges, volcanoes, basaltic seafloor crust. Enormous amounts of basalt have poured out of this world-encircling system of cracks. In the past 200 million years, enough magma has been released to lay down the crust of all the present seafloor.

Much of the volcanic heat on the seafloor is removed when cold seawater circulates in the fissures of the ocean-ridge volcanic system. Seawater that has been heated and enriched in dissolved minerals by its contact with magmas forms extremely hot

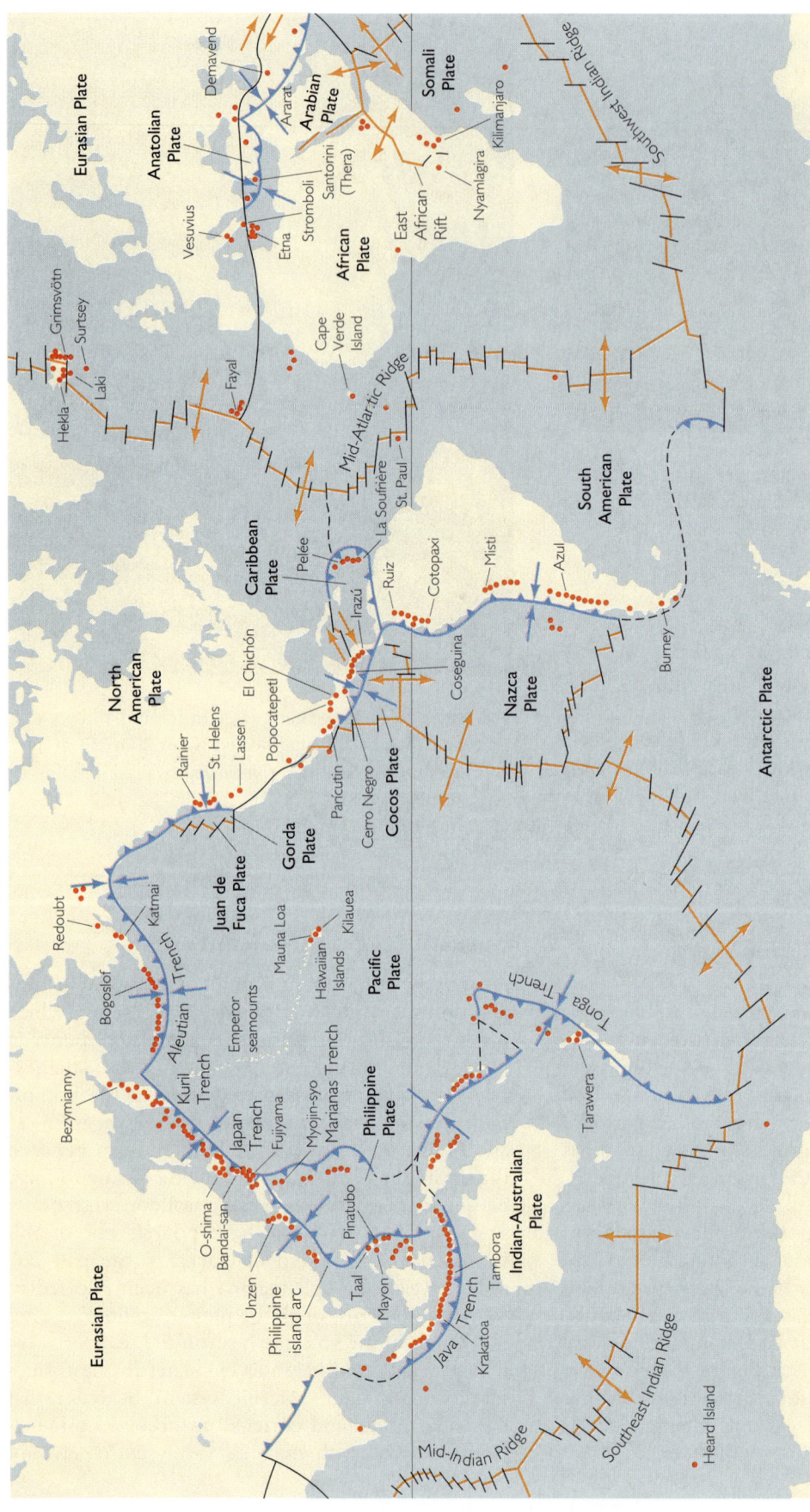

FIGURE 5.28 The active volcanoes of the world are not distributed randomly on Earth's surface. About 80 percent occur at boundaries where plates collide, 15 percent where plates separate, and the remaining few at intraplate hot spots. Convergent boundaries are shown in blue, divergent boundaries in orange. Black lines are transform faults. Active volcanoes are marked by red dots.

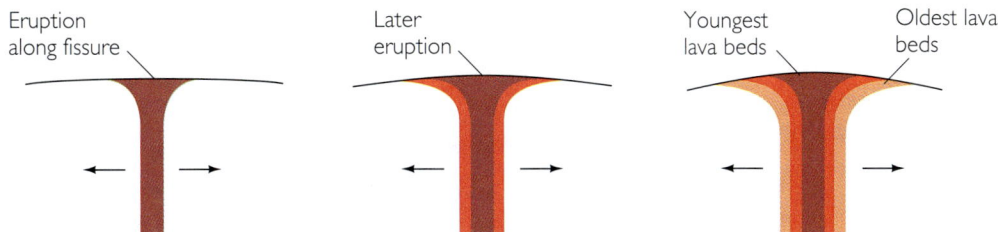

FIGURE 5.29 Iceland is an exposed part of the Mid-Atlantic Ridge. Repeated fissure eruptions and lateral spreading are the mechanisms of its growth.

(350°C) springs and smoking vents along the cracks. These sites are a major source of minerals including ores of zinc, copper, and iron (see Feature 17.2).

Iceland, one of the few exposed segments of the Mid-Atlantic Ridge, provides an unmatched opportunity to view the process of fissure eruption and seafloor spreading directly (see Figure 1.15). The island is composed mostly of basalt. Iceland is in a state of tension, literally being pulled apart—one half is moving eastward in the Eurasian Plate, the other westward in the North American Plate. Tensional forces cause fissures to develop, and magma flows in from below and overflows onto the surface. At the conclusion of each episode, the lava solidifies to form a nearly vertical dike in the fissure and nearly horizontal beds on the adjacent surfaces. With each new episode of lateral spreading, a new crack opens and another flow pours out over the old one (Figure 5.29). Iceland thus grows by repeated eruptions, primarily from long fissures but also from localized vents. Although the details may differ under water, it is likely that the seafloor crust grows in a similar fashion.

Convergence-Zone Volcanism

Scientists are now sorting out the many phenomena that occur where plates converge and subduction occurs. One of the most striking features is the chain of volcanoes that parallels convergent boundaries, whether their type is ocean-ocean or ocean-continent (Figure 5.30). Magmas that feed convergence-zone volcanoes are more varied than the basalts of mid-ocean ridge volcanism and range from mafic to felsic, that is, from basalts to andesites to rhyolites. Subduction provides several mechanisms to explain these observations.

PLATE SUBDUCTION AND CONVERGENCE-ZONE MAGMAS Because water lowers the melting temperature of rock (Chapter 4), water from the veneer of seafloor sediments atop the subducted slab can induce melting in the hot mantle above it. This process provides the basaltic magmas that feed convergence-zone volcanoes (see pages 85–86). In addition to basaltic magmas that derive from partial melting of the mantle, intermediate and more felsic magmas would form if sources were available to contribute silica and the other elements to the melt (Chapter 4; see Table 4.2 and Figure 4.8). In subduction zones, where the initially cool subducted slab heats up as it plunges into the hot mantle, such materials would derive from melting of the seafloor sediments and ocean crust atop the subducted ocean slab. Also, magmas rising from the mantle can invade and partially melt the felsic crust of an overriding continental plate. This process accounts for the more felsic kinds of magmas that feed volcanoes of an overriding continental plate margin. Thus the association of volcanism with subduction and the generation of magmas of different types is predicted by plate-tectonics theory.

VOLCANISM AT OCEAN–OCEAN CONVERGENCE In the case of ocean–ocean convergence, an arc of volcanic islands builds up from the seafloor of the overriding plate, mostly by the extrusion of basalts, occasionally andesites, and rarely rhyolites. The basalts probably derive from the asthenosphere above the descending plate. The more silicic andesites would occur when elements are added, derived from varying degrees of partial melting of the basaltic crust and the ocean-bottom sediments attached to the descending plate (as described in Chapter 4). The creation of the Aleutian and Mariana island arcs are prototypes of this process. The collision of two

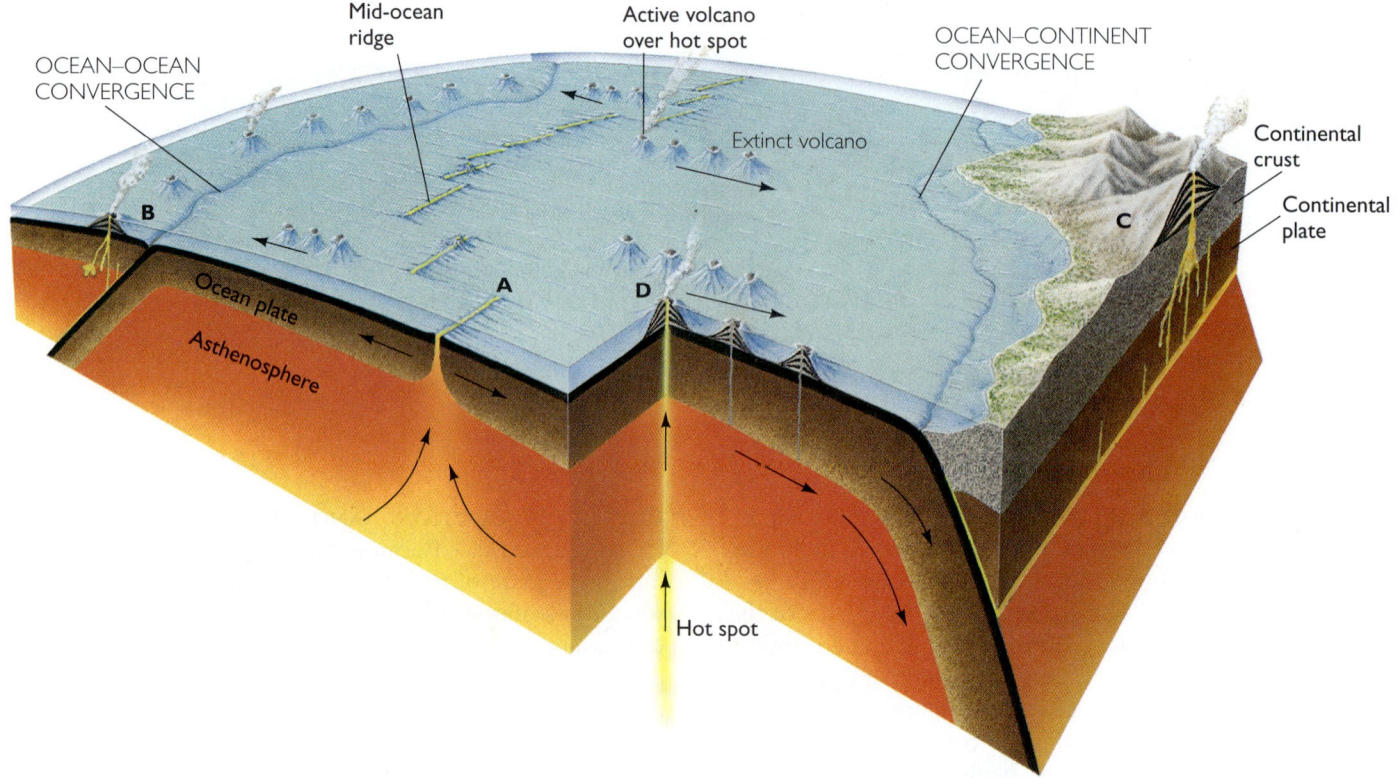

FIGURE 5.30 Volcanism is associated with plate tectonics. Plate separation at a mid-ocean ridge and partial melting of upwelling ultramafic mantle result in basaltic volcanism (A). At ocean–ocean convergent boundaries, magmas originating from partial melting of the mantle give rise to volcanic island arcs erupting mostly basaltic lavas (B). Magmas formed at ocean–continent convergences are mixtures of basalts from the mantle, remelted felsic continental crust, and materials melted off the top of the subducted plate. They give rise to volcanoes erupting andesitic and rhyolitic lavas (C). Plate motion over hot spots accounts for the creation of mid-plate chains of basaltic volcanic islands (D).

ocean plates is also responsible for Fujiyama, an andesitic volcanic cone, and Mount Pelée on the island of Martinique, from which viscous felsic lavas and explosive pyroclastic flows erupt.

VOLCANISM AT OCEAN–CONTINENT CONVERGENCE When a plate carrying a continent on its leading edge overrides an ocean plate, an arcuate volcanic mountain chain grows in the zone of collision near the continental margin (see Figure 5.30). The Andes mark the convergence boundary between South America and the Nazca Plate. Farther north, the subduction of the small Juan de Fuca Plate under the North American Plate gives rise to the volcanoes of the Cascade Range, which stretch from northern California to British Columbia. Mount St. Helens (see Feature 5.3, page 128) is among them. During a typical volcanic eruption, large quantities of ash and of andesitic lavas, and more rarely rhyolite, are ejected. The source is probably a mixture of basaltic magma rising from the mantle and remelted felsic continental crust through which it passes. Materials melted off the subducted slab could also contribute to the mixture.

Intraplate Volcanism

For many years, volcanism far from plate boundaries posed a problem for the theory of plate tectonics: it seemed to be an exception to the neat correlation of volcanism and plate boundaries. Take the Hawaiian Islands, in the middle of the Pacific Plate. This island chain begins with the active volcanoes on Hawaii and continues as a string of progressively older, extinct, eroded, and submerged volcanic ridges and mountains. Frequent large earthquakes do not occur along the Hawaiian chain—that is, it is essentially aseismic (without earthquakes)—so it is called an **aseismic ridge.**

Aseismic ridges of volcanic origin, which also occur elsewhere in the Pacific and in the other large oceans, were difficult to incorporate into the plate-tectonics framework until the concept of **hot spots** was introduced. Hot spots have also been proposed to explain some forms of volcanism within continents, far from plate boundaries. Yellowstone is an example. According to this hypothesis, illustrated in Figure 5.30, hot spots are the volcanic manifestations of jets or plumes of hot solid material that rise from deep within the mantle (perhaps even from the core-mantle boundary). When the plume reaches the lower pressures of shallow depths it begins to melt. The magma penetrates the lithosphere, and erupts at the surface. These columnar currents are thought to be fixed in the mantle and not to move with the lithospheric plates. As a result, the hot spot leaves a trail of extinct, progressively older volcanoes as the plate moves over it. The tracks of the extinct volcanoes that constitute the Hawaiian Islands and the Emperor seamount chain (shown in Figure 5.28) trace the motion of the Pacific Plate over a hot spot marked by the active volcanoes on Hawaii. The bend in the chain records a change in the direction of plate motion. (The satellite map of the seafloor in Feature 17.3 shows this hot spot trail clearly.)

If hot spots are indeed fixed in the mantle, the trail of volcanoes carried away from the hot spot provides a powerful method of measuring the velocity of plate motion. We could calculate the velocity, for example, by dividing the distance an extinct volcano has traveled from its hot spot by the age of that volcano's youngest volcanic rock. Data from deep-sea drilling along the Hawaiian and Emperor seamount chains have provided evidence consistent with the hypothesis that they originated over hot spots. Drill core samples confirm that the farther islands in these chains are from an active hot spot, the older they are. Calculations tell us that the plates are moving several centimeters a year and that the change in direction of plate motion occurred 40 million years ago.

The origin of fissure eruptions of basalt on continents, such as those that formed the Columbia River Plateau, and even larger lava plateaus in Brazil–Paraguay, India, and Siberia, is a subject of debate. Some geologists suggest that because of the immense amount of lava released (well over a million cubic kilometers in some cases) superplumes rising from deep within the mantle are the sources of the fissure eruptions that build the great continental lava plateaus. Others propose that fractures (of unknown origin) penetrated the continental lithosphere and that basaltic lavas, which represent partial melts of the underlying mantle, spurted rapidly to the surface without much contamination from the felsic crust. The volcanism that covered much of Siberia with lava is of special interest because it occurred at the end of the Permian some 250 million years ago and may have caused the greatest mass extinction of species in the geological record (see Chapter 24).

Fissure eruptions that mark the initial stages of continental rifting and the opening of a new ocean can be documented in several parts of the world. For example, basalt is found in the rift valleys of East Africa (see Figure 5.28)—a feature that some geologists interpret as signs of a breakup of Africa that was never completed.

Volcanism and Human Affairs

Of the many volcanoes that have affected communities worldwide, one made a particularly powerful impact on Western civilization. Teams of archeologists and marine geologists have pieced together the story of the demise of Thera (formerly Santorini), a volcanic island in the Aegean Sea. The eruption of Thera in about 1623 B.C. appears to have been far more violent than that of Krakatoa. The center of the island collapsed, forming a caldera visible today as a lagoon about 60 km in circumference and as much as 500 m deep, with two small active volcanoes in the center. The lagoon is rimmed by two crescent-shaped islands known for their wine exports and scenic beauty and still subject to destructive earthquakes. The resultant volcanic debris and tsunami from this ancient catastrophe destroyed dozens of coastal settlements over a large part of the eastern Mediterranean. Some scientists have attributed the mysterious disappearance of the Minoan civilization to this cataclysm. And it is thought that the legend of the lost continent of Atlantis may have its origin in the land collapse that accompanied the eruption. The course of history was probably changed by this one volcanic event. It could happen again.

The date is known from a layer of volcanic ash transported to Greenland from Santorini. The ash layer was found in 1994 in an ice core extracted from deep within the Greenland ice sheet. Annual cycles of snow deposition can be counted in a core as rings in a tree are counted to determine its age. The ice core goes back 7000 years and provides evidence of some 400 ancient volcanic eruptions.

Can volcanic eruptions be predicted? To some extent they can (Feature 5.2)—fortunately for us all, because there are about 100 high-risk volcanoes in the world, and some 50 erupt each year. Certainly with our growing understanding of volcanism we can improve the terrible record of the past. Over the last 500 years alone some 200,000 people have been killed by volcanic eruptions.

Reducing the Risks of Hazardous Volcanoes

Of Earth's 500 to 600 active volcanoes, one out of six has claimed human lives. Volcanoes kill people and damage property by edifice collapse, explosive blasts, ash falls, lethal gas release, lava flows, and mudflows or lahars.

Scientists monitoring Mount St. Helens (Feature 5.3) and Mount Pinatubo were able to issue warnings of imminent major eruptions (Figure 5.31).

Government infrastructures were in place to evaluate the warnings and to issue and enforce evacuation orders. In the case of Pinatubo, the warning was issued a few days before the cataclysmic eruption on May 17, 1991. A quarter of a million people were evacuated, including some 16,000 residents of the nearby U.S. Clark Air Force Base (since permanently abandoned). Tens of thousands of lives were saved from the lahars that destroyed everything in their path. Casualties were limited to the few who disregarded the order. In 1994, 30,000 residents of Rabaul, Papua New Guinea, were successfully evacuated by land and sea hours before two volcanoes on either side of the town erupted, destroying or damaging most of it. Many owe their lives to the government for conducting evacuation drills and to scientists at the local volcano observatory who issued a warning when their seismographs recorded the ground tremor that signaled magma moving toward the surface.

Contrast these success stories with the tragedy at Nevado del Ruiz in Colombia in 1985. Scientists knew this volcano to be dangerous and were prepared to issue warnings, but no evacuation procedure was in place. As a result, 25,000 lives were lost in the lahars triggered by a minor eruption.

Volcanology has progressed to the point that we can identify the world's dangerous volcanoes and characterize their potential hazards from deposits laid down in earlier eruptions. These hazard assess-

FIGURE 5.31 Scientists carried by helicopter into the caldera of Mount Pinatubo, Philippines, collecting samples of gas, water, and volcanic debris. *(Roger Ressmeyer/Corbis.)*

5.2 TECHNOLOGY AND EARTH

Kilauea: Monitoring a Volcano

The giant shield volcano Mauna Loa and the smaller Kilauea on its eastern flank make up the southern half of the island of Hawaii. Because the U.S. Geological Survey operates a volcano observatory on the rim of Kilauea Caldera, this volcano, which is actively erupting, is perhaps the best studied in the world. What has been learned from it has profoundly influenced our notions of volcanic processes.

A modern network of instruments and laboratory facilities is used to track the movement of magma within the volcano and the changing chemistry of the erupting lavas and gases. Seismographs, which measure movements within the Earth, detect and locate the small earthquakes that are often correlated with movements of magma. Seismographs can locate earthquakes as deep as 55 km beneath Kilauea. Such quakes often mark the entrance of magma into the channels leading from the asthenosphere through the lithosphere to the Earth's surface. The upward migration of the magma can be followed over a period of months because seismic disturbances occur progressively nearer the surface as the magma rises. Tiltmeters, which measure tilting of the ground, indicate when the volcano begins to swell as the rising magma fills a magma chamber not far below the summit.

The first sign that an outbreak of lava is imminent is a swarm of small earthquakes, thousands of them, signifying that the magma is splitting rock as it forces its way to the surface. Very often, geologists know where the eruption will occur from the location of the earthquakes and changes in their pattern. In January 1960, for example, U.S. Geological Survey scientists detected a swarm of earthquakes not far from the village of Kapoho, on the flank of Kilauea. As they expected, an eruption broke out, destroying Kapoho but causing no casualties because the village had been evacuated. A new landscape was created as the lava flowed to the sea. Twenty-foot walls were built in a futile attempt to divert the lava and save a seashore community. When it was all over, the tiltmeters showed that the volcano had deflated, signifying that the magma chamber below had been drained in the Kapoho eruption. The cycle is repeated every few years.

Destruction of property engulfed by lava flow in the May 1990 eruption of Kilauea volcano, Hawaii. *(James Cachero/Sygma.)*

5.3 INTERPRETING THE EARTH

Mount St. Helens: Dangerous but Predictable

Long before Mount St. Helens erupted in the spring of 1980, geologists knew it to be the most active and explosive volcano in the contiguous United States. They could piece together a 4500-year history of destructive lava flows, hot pyroclastic flows, lahars, and distant ash falls by examining the geologic record. Monitoring efforts were intense. Beginning on March 20, 1980, a series of small to moderate earthquakes under the volcano signaled the start of a new eruptive phase after 123 years of dormancy. The earthquakes moved the U.S. Geological Survey (USGS) to issue a formal hazard alert. The first outburst of ash and steam erupted from a newly opened crater on the summit one week later. In April the seismic tremors increased, indicating that magma was moving beneath the summit, and an ominous swelling of the northeastern flank was noticed. The USGS issued a more serious warning, and people were ordered out of the vicinity.

On May 18 the climactic eruption began abruptly. A large earthquake apparently triggered the collapse of the north side of the mountain, loosening a massive landslide, the largest in recorded world history. As a huge flow of debris plummeted down the mountain, gas and steam under high pressure were released in a tremendous lateral blast that blew out the northern flank of the mountain. USGS geologist David A. Johnston was monitoring the volcano from his observation post 8 km to the north. He must have seen the advancing blast wave before he radioed his last message: "Vancouver, Vancouver, this is it!" A northward-directed jet of superheated (500°C) ash, gas, and steam roared out of the breach with hurricane force, devastating a zone 20 km outward from the volcano and 30 km wide. A vertical eruption sent an ash plume 25 km into the sky, twice as high as a commercial jet flies. The ash cloud drifted to

Mount St. Helens before and after the cataclysmic eruptions of May 1980.
(Before: Emil Muench/Photo Researchers. After: David Weintraub/Photo Researchers.)

ments can be used to guide zoning regulations to restrict land use—the most effective measure to reduce casualties. Instrumented monitoring (as described in Feature 5.2) can detect signals such as earthquakes, swelling of the volcano, and gas emissions that warn of impending eruptions. People at risk can be evacuated if the authorities are organized and prepared. Volcanic eruptions cannot be prevented, but their catastrophic effects can be significantly reduced by a combination of science and enlightened public policy.

As a part of a United Nations program—the International Decade of Natural Disaster Reduction—leading volcanologists have designated 15 active the east and northeast with the prevailing winds, bringing darkness at noon to an area 250 km to the east and depositing ash up to 10 cm deep on much of Washington, northern Idaho, and western Montana. The energy of the blast was equivalent to about 25 million tons of TNT. The volcano's summit was destroyed, its elevation reduced by 400 m, its northern flank disappeared. In effect, the mountain was "hollowed out."

Local devastation was spectacular. Within an inner blast zone extending 10 km, the thick forest was denuded and buried under several meters of pyroclastic debris. Beyond this zone, out to 20 km, trees were stripped of their branches and blown over like broken matchsticks aligned radially away from the volcano. As far as 26 km away, the hot blast was so intense that it overturned a truck and melted its plastic parts. Some fishermen were severely burned and survived only by jumping into a river. More than 60 other people were killed by the blast and its effects.

A lahar formed when the landslide and pyroclastic debris—fluidized by groundwater, melted snow, and glacial ice—flowed 28 km down the valley of the Toutle River. The valley bottom was filled to a depth of 60 m. Beyond this debris pile, muddy water flowed into the Columbia River, where sediments clogged the ship channel and stranded many vessels in Portland. Mount St. Helens may go on erupting for 20 years or more until its present episode of activity comes to an end.

Although much of the devastated area is still barren, after a decade almost 20 percent of the surface showed evidence of revegetation: native species of grass, legumes, and young trees are beginning to come back.

TABLE 5.1

15 VOLCANOES CHOSEN BY THE UNITED NATIONS FOR STUDY, 1995–2000

VOLCANO	COUNTRY
Colima	Mexico
Etna	Italy
Galeras	Colombia
Mauna Loa	United States
Merapi	Indonesia
Niragongo	Zaire
Rainier	United States
Sakurajima	Japan
Santa Maria/Santiaguito	Guatemala
Santorini	Greece
Taal	Philippines
Teide	Spain (Canary Islands)
Ulawun	Papua New Guinea
Unzen	Japan
Vesuvius	Italy

volcanoes for special study for the period 1995–2000 (Table 5.1). Many of these are among the most dangerous in the world. The goals of the project are to understand all volcanoes better, to evaluate the hazards posed by these particular volcanoes, to improve the prediction of explosive eruptions, and to prepare the public for crises that may develop.

Volcanologists who study active volcanoes are motivated to understand them so that warnings of eruptions can be made more reliable. Their work is hazardous—nine have lost their lives since 1991. In one case a team of seven volcanologists gathering data in the crater of Galeras volcano in the Colombian Andes was caught in a pyroclastic eruption. Six of the seven members lost their lives in the ash and incandescent boulders that exploded from the crater. The lone survivor, Professor Stanley Williams of Arizona State University, is now at work developing an instrument that can analyze volcanic gases in the crater of an active volcano and transmit this information to scientists at a safe distance. An impending eruption might be predicted from the kinds of gases that are emitted.

Can volcanic eruptions be controlled? Not likely, although in special circumstances and on a small scale the damage can be reduced. Perhaps the most successful attempt to control volcanic activity was made on the Icelandic island of Heimaey in January 1973. By spraying advancing lava with seawater, Icelanders cooled and slowed the flow, preventing the lava from blocking the port entrance and saving some homes from destruction.

In the years ahead, the best policy for protecting the public will be the establishment of more warning and evacuation systems and more rigorous restriction of settlements in potentially dangerous locations. But even these precautions may not help. Dormant or long-extinct volcanoes can come to life suddenly—as Vesuvius and St. Helens did after hundreds of years. (Some potentially dangerous volcanoes in the United States and Canada are identified in Figure 5.32.) An even more difficult problem of prediction is posed by eruptions like that of Parícutin, which rose up with little warning from a small hole in a Mexican cornfield in 1943. Whole towns were quickly buried by ash and lava as the new volcano grew by repeated eruptions. Learning how to sense the movements of deep lava in relation to possible new outlets to the surface is a real challenge for geologists.

Reaping the Benefits of Volcanoes

We have seen something of the beauty of volcanoes and also something of their destructiveness. Volcanoes contribute to our well-being in many ways. In Chapter 1 we mentioned that the atmosphere and the oceans may have originated in volcanic episodes of the distant past. Soils derived from volcanic materials are exceptionally fertile because of the mineral nutrients they contain. Volcanic rock, gases, and steam are also sources of important industrial materials and chemicals, such as pumice, boric acid, ammonia, sulfur, carbon dioxide, and some metals. Seawater circulating through fissures in the ocean-ridge volcanic system is a major factor in the formation of ores.

Thermal energy from volcanism is being harnessed in more and more places. Most of the houses

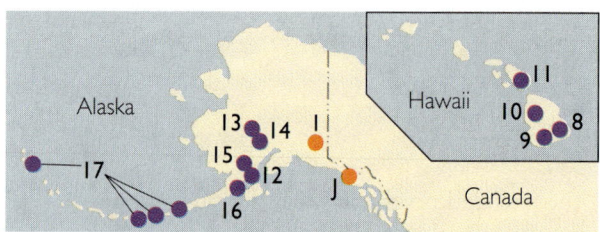
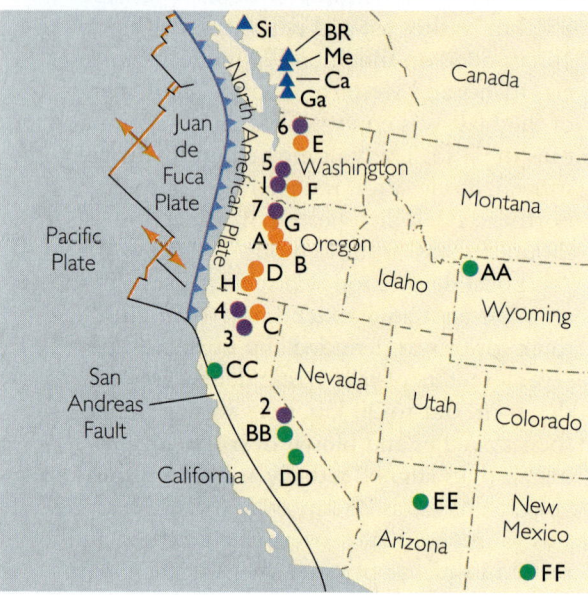

- U.S. volcanoes that have short-term eruption periodicities (100–200 years or less), or have erupted in the past 200–300 years, or both:

Cascades	Hawaii	Alaska
1 Mount St. Helens	8 Kilauea	12 Augustine volcano
2 Mono-Inyo craters	9 Mauna Loa	13 Redoubt volcano
3 Lassen Peak	10 Hualalai	14 Mount Spurr
4 Mount Shasta	11 Haleakala	15 Iliamna volcano
5 Mount Rainier		16 Katmai volcano
6 Mount Baker		17 Aleutian volcanoes
7 Mount Hood		

- U.S. volcanoes that appear to have eruption periodicities of 1000 years or greater and last erupted 1000 years or more ago:

Cascades	Alaska
A Three Sisters	I Mount Wrangell
B Newberry volcano	J Mount Edgecumbe
C Medicine Lake volcano	
D Crater Lake (Mount Mazama)	
E Glacier Peak	
F Mount Adams	
G Mount Jefferson	
H Mount McLoughlin	

- U.S. volcanoes that last erupted more than 10,000 years ago, but beneath which exist large, shallow bodies of magma that are capable of producing exceedingly destructive eruptions:

AA Yellowstone Caldera	DD Coso volcanoes
BB Long Valley Caldera	EE San Francisco Peak
CC Clear Lake volcanoes	FF Socorro

▲ Danger classifications are not available for Canadian volcanoes:
Si Silverthrone BR Bridge River Me Meagher Mountain
Ca Mount Cayley Ga Mount Garibaldi

FIGURE 5.32 Locations of potentially hazardous volcanoes in the United States and Canada. Volcanoes within each U.S. group are listed in the order of declining probable cause for concern, subject to revision as studies progress (danger classifications are not available for Canadian volcanoes). Note the relationship between the Cascade volcanoes, which extend from northern California to British Columbia, and the subduction plate boundary between the North American Plate and the Juan de Fuca Plate. *(After R. A. Bailey, P. R. Beauchemin, F. P. Kapinos, and D. W. Klick, USGS.)*

FIGURE **5.33**
Crops growing on fertile volcanic soils in the Canary Islands. *(Cesar Lucas/ The Image Bank.)*

in Reykjavík, Iceland, are heated by hot water tapped from volcanic springs. Geothermal steam, originating in water heated by contact with hot volcanic rocks below the surface, is exploited as a source of energy for the production of electricity in Italy, New Zealand, the United States, Mexico, Japan, and the former Soviet Union. Figure 5.33 shows the agricultural benefits of volcanic soils.

Summary

Why does volcanism occur? Volcanism occurs when molten rock inside the Earth rises buoyantly to the surface because it is less dense than surrounding rock. In effect, the melt is squeezed up by the weight of the overlying layers.

What are the three major categories of lava? Lavas are classified as felsic (rhyolite), intermediate (andesite), or mafic (basalt), on the basis of the decreasing amounts of silica and the increasing amounts of magnesium and iron they contain. The chemical composition and gas content of lava are important factors in the form an eruption takes.

How are the structure and terrain of a volcano related to the kind of lava it emits and the style of its eruption? Basalt can be highly fluid. On continents it can erupt from fissures and flow out in thin sheets to build a lava plateau. A shield volcano grows from repeated eruptions of basalt from vents. Silicic magma is more viscous and, when charged with gas, tends to erupt explosively. The resultant pyroclastic debris may pile up into a cinder cone or cover an extensive area with ash-flow sheets. A stratovolcano is built of alternating layers of lava flows and pyroclastic deposits. The rapid ejection of magma from a magma chamber a few kilometers below the surface, followed by collapse of the chamber's roof, results in a large surface depression, or caldera. Giant resurgent calderas are among the most destructive natural cataclysms.

How is volcanism related to plate tectonics? The ocean crust forms from basaltic magma that rises from the asthenosphere into fissures of the ocean ridge–rift system where plates separate. Basaltic, andesitic (intermediate), and rhyolitic (felsic) lavas tend to erupt in convergent zones. The basalts derive from partial melting of the mantle above the subducted plate, induced by water streaming off subducted seafloor sediments. Basalts are typical of volcanic islands found at ocean–ocean plate convergences. Andesites and rhyolites are more commonly found in the volcanic belts of ocean–continent convergent plate boundaries. The addition to basaltic magma of silica and other elements derived from remelting of felsic continental crust or from melting of seafloor sediments and crust atop the downgoing slab can produce andesites and rhyolites. Within plates volcanism may occur above hot spots, which are manifestations of plumes of hot material that rise from deep in the mantle.

What are some beneficial effects of volcanism? Over the course of Earth's evolution, volcanic eruptions released the water and gases that formed the oceans and atmosphere. Geothermal heat drawn from areas of recent volcanism is of growing importance as a source of energy. An important ore-forming process occurs when groundwater circulates around buried magma or seawater circulates through ocean floor rifts.

Key Terms and Concepts

volcanism (p. 106)
volcano (p. 106)
lava (p. 106)
pahoehoe (p. 108)
aa (p. 108)
pillow lava (p. 108)
ash (p. 109)
volcanic tuff (p. 110)
volcanic breccia (p. 110)
pyroclastic flow (p. 110)

central vent (p. 111)
shield volcano (p. 111)
volcanic dome (p. 111)
cinder cone (p. 112)
composite volcano/
  stratovolcano (p. 112)
crater (p. 113)
caldera (p. 114)
phreatic explosion (p. 115)
diatreme (p. 117)

fissure eruption (p. 117)
flood basalts (p. 118)
ash-flow deposit (p. 118)
lahar (p. 119)
fumarole (p. 120)
aseismic ridge (p. 125)
hot spot (p. 125)

Exercises

1. The asthenosphere has been identified as a major source of magma. Why? What forces magma to rise to the surface?

2. What is the difference between magma and lava? Give examples of types of volcanic rocks and their coarse-grained, intrusive counterparts.

3. Describe the principal styles of eruptions and the deposits and landforms each style produces.

4. The accompanying photograph shows the remains of a building in a village 5 km south of El Chichón volcano, in southeastern Mexico. The village was destroyed when El Chichón erupted in April 1982. From the debris and the bent reinforcing rods evident in the photograph, what can you conclude about the nature of the flow, its force, and its direction?

5. What is the association between plate boundaries and volcanism? Can the eruptive style and composition of volcanic deposits be correlated with plate boundaries?

6. Under what circumstances do lahars occur? Hot springs? Ash-flow deposits?

7. Name the most dangerous features of volcanoes.

8. What signals an impending eruption?

Thought Questions

1. What public policy initiatives do the eruptions of Mount St. Helens, Mount Pinutabo, and Nevado del Ruiz suggest should be undertaken in such areas as zoning, land use, insurance, warning systems, and public education?

2. Do a risk-benefit analysis of volcanoes—that is, tabulate their dangers and their contributions to humankind—and decide whether you would prefer an Earth with or without them.

3. What have we learned about the Earth's interior from volcanoes?

Suggested Readings

A.G.U. Special Report. 1992. *Volcanism and Climatic Change.* Washington, D.C.: American Geophysical Union.

Decker, R. W., and B. Decker. 1989. *Volcanoes,* rev. and updated ed. New York: W. H. Freeman.

Dvorak, John J., Carl Johnson, and Robert I. Tilling. 1982. Dynamics of Kilauea volcano. *Scientific American* (August):46–53.

Edmond, John M., and Karen L. Von Damm. 1992. Hydrothermal activity in the deep sea. *Oceanus* (Spring):74–81.

Francis, Peter. 1983. Giant volcanic calderas. *Scientific American* (June):60–70.

Heiken, G. 1979. Pyroclastic flow deposits. *American Scientist* 67:564–571.

Krakaner, Jon. 1996. Geologists worry about dangers of living "under the volcano." *Smithsonian,* 33–125.

McPhee, John. 1990. Cooling the lava. In *The Control of Nature.* New York: Farrar, Straus & Giroux.

National Research Council. 1994. *Mount Rainier, Active Cascade Volcano.* Washington, D.C.: National Academy Press.

Simkin, T., L. Siebert, L. McClelland, D. Bridge, C. Newhall, and J. H. Latter. 1981. *Volcanoes of the World.* New York: Academic Press.

Tilling, Robert I. 1989. Volcanic hazards and their mitigation: Progress and problems. *Reviews of Geophysics* 27 (no. 2):237–269.

Vink, Gregory E., and W. Jason Morgan. 1985. The Earth's hot spots. *Scientific American* (April):50–57.

White, Robert S., and Dan P. McKenzie. 1989. Volcanism at rifts. *Scientific American* (July):62–72.

Wright, Thomas L., and Thomas C. Pierson. 1992. *Living with Volcanoes.* U.S. Geological Survey Circular 1073. Washington, D.C.: U.S. Government Printing Office.

Internet Sources

Cascades Volcano Observatory
http://vulcan.wr.usgs.gov/home.html
This U.S. Geological Survey laboratory is the major source of information about the Mount St. Helens eruption and monitoring of the Cascade volcanoes. The site provides links to images, a data archive, and information on preparing for a Cascade volcano eruption.

The Electronic Volcano
http://www.dartmouth.edu/pages/rox/volcanoes/elecvolc.html
This site is a "front door" for links to data sets and images of volcanoes around the world.

Hawaii's Center for Volcanology
http://www.soest.hawaii.edu/GG/hcv.html
Maintained by the School of Ocean and Earth Science and Technology of the University of Hawaii (Manoa), this site includes images and data for past and current eruptions in the Hawaiian Islands.

Volcanoes
http://www.geo.mtu.edu/volcanoes/
Michigan Technological University maintains this extensive site with reference maps, recent and ongoing activity, information about volcanic hazards mitigation, a glossary, and even volcano humor.

VolcanoWorld
http://volcano.und.nodak.edu/vw.html
This site, at the University of North Dakota, is a "front door" for almost everything available on volcanoes. Features include Today in Volcano History, What's Erupting Now? (providing a map and links to each eruption with images and a glossary), Ask a Volcanologist, Volcanic Peaks and Monuments, and Volcanoes of the World.

Japan's Volcano Research Center
http://hakone.eri.u-tokyo.ac.jp/vrc/VRC.html
Current volcanic activity in Japan is the focus of this site at the University of Tokyo. Current reports are maintained on the major Japanese volcanoes, including Fuji and Unzen.

THE GEOLOGIC TIME SCALE

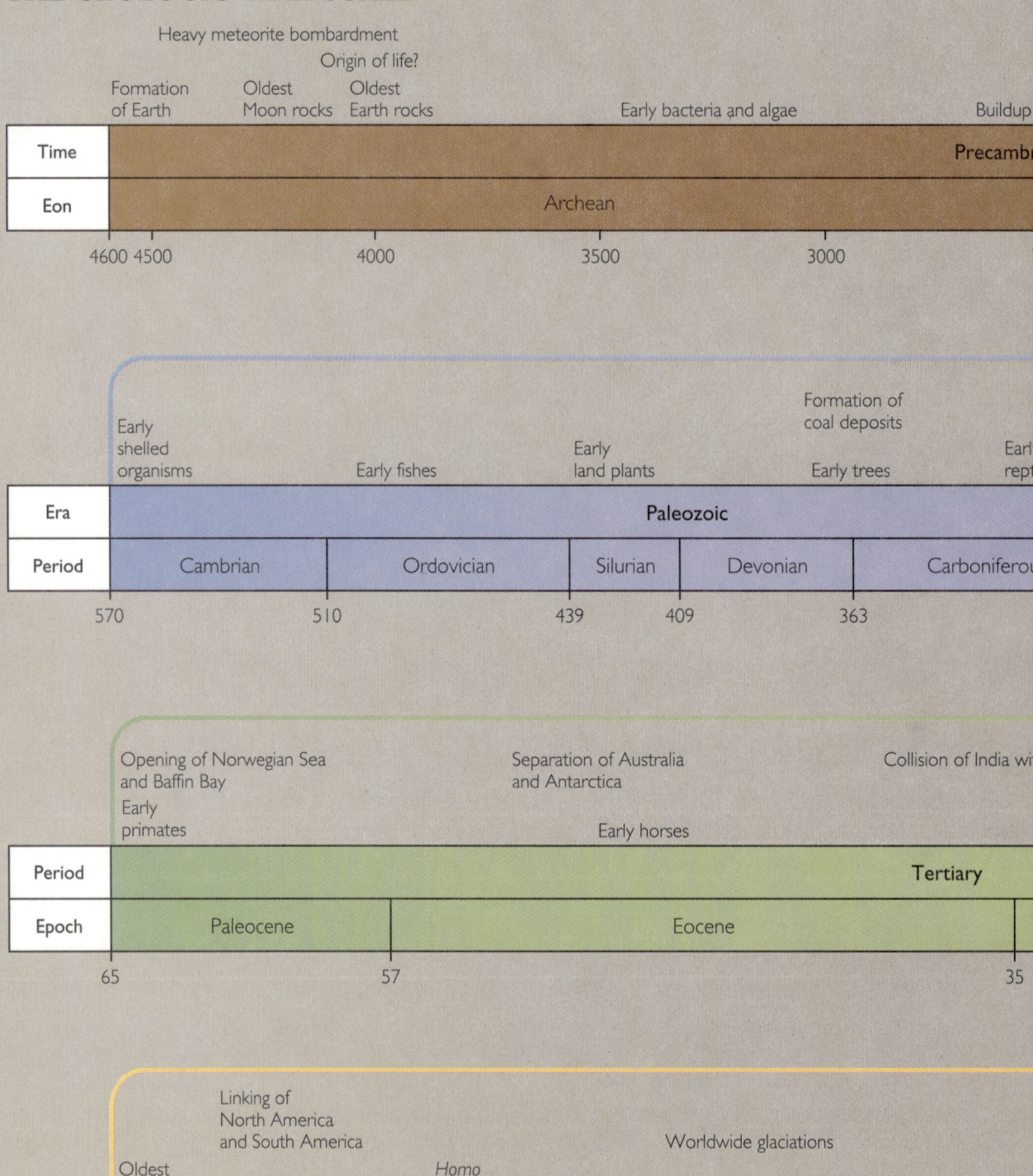

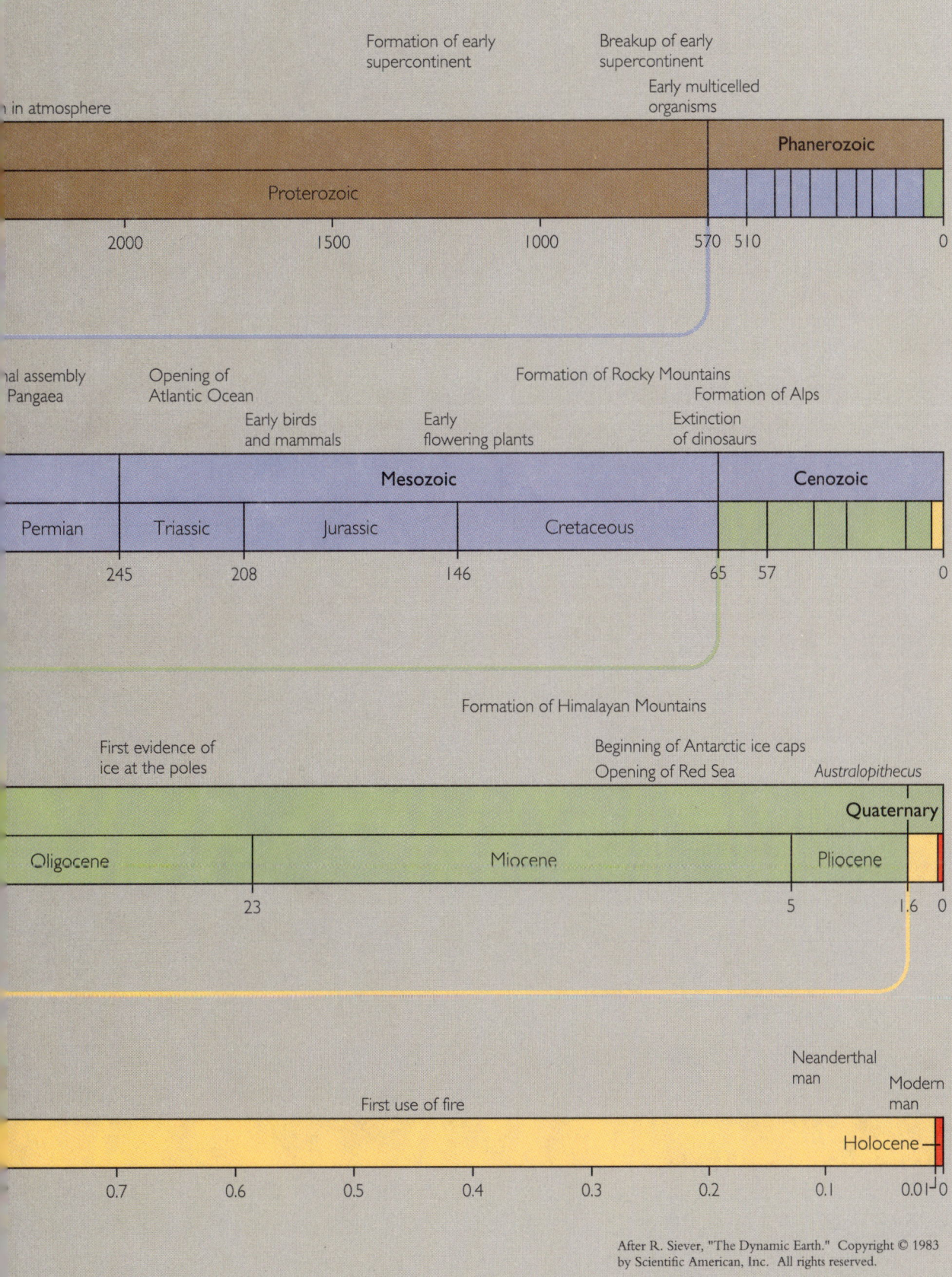

After R. Siever, "The Dynamic Earth." Copyright © 1983 by Scientific American, Inc. All rights reserved.

Updated from W. B. Harland et al., *Geologic Timescale*, 1989.